UNUSUALLY SPECIAL RELATIVITY

UNUSUALLY
SPECIAL
RELATIVITY

V

Andrzej Dragan

University of Warsaw, Poland &
National University of Singapore, Singapore

 **World Scientific**

NEW JERSEY · LONDON · SINGAPORE · BEIJING · SHANGHAI · HONG KONG · TAIPEI · CHENNAI · TOKYO

Published by

World Scientific Publishing Europe Ltd.

57 Shelton Street, Covent Garden, London WC2H 9HE

Head office: 5 Toh Tuck Link, Singapore 596224

USA office: 27 Warren Street, Suite 401-402, Hackensack, NJ 07601

Library of Congress Cataloging-in-Publication Data
Names: Dragan, Andrzej, author.
Title: Unusually special relativity / Andrzej Dragan, University of Warsaw,
    Poland & National University of Singapore, Singapore.
Description: London ; Hackensack, NJ : World Scientific Publishing Europe Ltd., [2022] |
    Includes bibliographical references and index.
Identifiers: LCCN 2021026274 (print) | LCCN 2021026275 (ebook) |
    ISBN 9781800610804 (hardcover) | ISBN 9781800610880 (paperback) |
    ISBN 9781800610811 (ebook for institutions) | ISBN 9781800610828 (ebook for individuals)
Subjects: LCSH: Special relativity (Physics)
Classification: LCC QC173.65 .D726 2022 (print) | LCC QC173.65 (ebook) | DDC 530.11--dc23
LC record available at https://lccn.loc.gov/2021026274
LC ebook record available at https://lccn.loc.gov/2021026275

British Library Cataloguing-in-Publication Data
A catalogue record for this book is available from the British Library.

For any available supplementary material, please visit
https://www.worldscientific.com/worldscibooks/10.1142/Q0319#t=suppl

Desk Editors: Jayanthi Muthuswamy/Michael Beale/Shi Ying Koe

Typeset by Stallion Press
Email: enquiries@stallionpress.com

Printed in Singapore

Preface (Before Consuming)

When I was a student, I cooked up an early draft of this book. I was starving for a comprehensive and basic course on relativity in my syllabus! Then over the next 20 years I amused my own students (and myself) by dishing out samples from the rough contents of this draft. After countless in vivo tests and gradual improvements to the book's recipe, dinner is now ready to be served to a wider audience. And I trust you will find it to be a tasty meal, as special relativity is not only unexpectedly simple and intuitive, but also incredibly interesting!

To understand this book, advanced knowledge of physics is not required. The only thing you need is the ability to focus and recalibrate your common sense, as relativity can be a shock to your perception of "reality". To ease the mathematical burden I have reduced the tools needed to assimilate and process the material to the bare minimum. As a result, most topics can be grasped with basic understanding of high-school algebra. This is in stark contrast to the existing relativity literature, in which the overuse of complex mathematics quickly confuses the reader and clouds the comprehension. And what we can't explain to our own grannies, we probably just don't understand.

A delicious example of this is the phenomenon of the Thomas precession. It is typically skipped in most relativity textbooks for being too technical. And when it is considered, it is often with an intellectual repertoire involving "associative-commutative groupoids", "gyro-group vector spaces", "holonomy transformation groups", "Clifford algebras", "tetrad formalism" — enough to turn the most stoic of brains into mush! In my approach, the Thomas precession is derived within three lines using basic, high-school vector calculus. Being a fan of education and not necessarily "educationism", I trust that my book will serve as a nutritious appetiser to anyone beginning their first encounter with relativity. Those with a hunger for a more formalised, sophisticated, and mathematised approach are welcome to satiate themselves by turning to another book from the rich repertoire on the topic.

We will derive Lorentz transformations using three completely different methods. We will discuss in detail all known and unknown apparent paradoxes and riddles of special relativity, providing insights into these problems from many opposing angles and enabling a deeper understanding of the subject. Within this book we will go well beyond the basics, eventually touching on topics typically considered to be "more advanced", such as non-inertial accelerated reference frames, static black holes, and even white holes. We will step into what it means to be quantum and how that is connected to relativity. All of this is presented as simply and comprehensibly as possible, without skipping the essence. There will be orthodox verses and there will be superluminal apocrypha. And for the inevitable errors that slipped in unnoticed, may the Devil carry me bareback kicking and screaming into the abyss of disgrace. Smacznego![a]

Big thanks to Sam Akina, Matthew Halstead and Jules Jones for their invaluable help!

---

[a]Bon appétit!

About the Author

Photographer:
Paweł Fabjański

Andrzej Dragan is a professor of physics at University of Warsaw and a visiting professor at National University of Singapore, where he is leading a research group on relativistic quantum information. A former research fellow at Imperial College London and University of Nottingham, Professor Dragan is also the recipient of multiple awards and scholarships, including ones from European Science Foundation, State Committee for Scientific Research, Award for the Young Scholars from the Foundation for Polish Science, and the Minister of Education Award for Outstanding Scholars. He is also a former Scientific Secretary and a longtime member of the Head Committee of the Physics Olympiad.

Also known for the photographic "Dragan effect", Professor Dragan was named Photographer of the Year by the Digital Camera Magazine UK, nominated to the Golden Lion award at the Cannes advertising festival, awarded Best in Show from the Creative Review UK and declared winner of the London Fashion Film Festival, to name just a few, for his short films. He has never tasted coffee.

Scan the QR code to access the author's collection of Youtube videos that explain the fundamental concepts of physics described in *Unusually Special Relativity*.

Contents

Chapter 1

Let There Be (The Speed of) Light

1.1 Lorentz Transformation *à la* Minkowski

Hermann Minkowski, who was Albert Einstein's maths teacher, once had a clever idea. The whole story started with a peculiar fact known from experiments: light in a vacuum always moves at the same speed c. What do we mean by "always"? A more precise statement would be that light moves at the same speed relative to all inertial observers, i.e. those observers that do not accelerate (an issue that we will return to). To illustrate how weird that is, let's consider a simple example in which a ray of light leaves a headlight mounted by a Witch on her flying broom. This ray always departs the headlamp at the speed c. The striking fact is that, the ray approaches a stationary traffic cop at the same speed c, no matter how fast the Witch is flying. It would still be c even if the cop was moving with any speed. One may think that the light should approach the resting cop at the speed equal to $c + V$, where V is the speed of the Witch. However, nature does not care what we think. If the world was the way we think it is — it would be completely different.

So, let us try to be a little more precise. Suppose that our ray is moving between two points, **A** and **B**, separated in space by coordinate distances Δx, Δy, Δz and the travel time is Δt. That allows us to write a simple equation of motion: (travelled distance) = (speed)·

Figure 1.1: Two observers moving with respect to each other with a relative velocity V along the x axis. Origins of their reference frames coincide at $t = t' = 0$. The same notation (primed and unprimed coordinates) will be used throughout the book unless it is be stated otherwise.

(time lapsed), or

$$\sqrt{\Delta x^2 + \Delta y^2 + \Delta z^2} = c\Delta t. \tag{1.1}$$

According to all known experimental results, the speed c is identical for all inertial observers watching the light move. So, if equation (1.1) describes the situation from the point of view of the resting traffic cop, then an analogous equation can be written by the Witch flying on her broomstick with velocity V. Let us denote the coordinates used by the latter with primed symbols — as shown in Fig. 1.1.

Our goal will be to determine a coordinate transformation between the two considered reference frames, i.e. to relate the time and space coordinates of the first and the second observer [1]. We will only assume that such a transformation does not distinguish any point in time or space, which means that the coordinate transformation must be linear. The constancy of the speed of light means that if the first observer witnesses the equality $0 = c^2\Delta t^2 - \Delta x^2 - \Delta y^2 - \Delta z^2$ then according to the second observer, the analogous equality must be satisfied: $0 = c^2\Delta t'^2 - \Delta x'^2 - \Delta y'^2 - \Delta z'^2$. So, in other words, the two polynomials: $c^2\Delta t^2 - \Delta x^2 - \Delta y^2 - \Delta z^2$ and $c^2\Delta t'^2 - \Delta x'^2 - \Delta y'^2 - \Delta z'^2$ have all their zeros in common. In this case it can be shown that these two polynomials must be proportional to each other:

$$c^2\Delta t^2 - \Delta x^2 - \Delta y^2 - \Delta z^2 = K\left(c^2\Delta t'^2 - \Delta x'^2 - \Delta y'^2 - \Delta z'^2\right), \tag{1.2}$$

with some unknown proportionality factor K depending continuously on the relative velocity of the observers, V. Since both polynomials involved are quadratic and therefore do not depend on the orientation of the axes, the proportionality factor can only depend on the value of the velocity, not its direction. However, starting from the primed reference frame and going to the unprimed one, we could obtain the inverse relation:

$$c^2 \Delta t'^2 - \Delta x'^2 - \Delta y'^2 - \Delta z'^2 = K \left(c^2 \Delta t^2 - \Delta x^2 - \Delta y^2 - \Delta z^2 \right), \quad (1.3)$$

which upon substituting into (1.2) leads to the condition $K(V)^2 = 1$. Since for $V = 0$ we have $K = 1$ then the continuous function must satisfy $K = 1$ for all values of V.

All these considerations can be summarised by introducing a new temporal variable $\tau \equiv ict$ (where $i = \sqrt{-1}$), a mysterious "imaginary time", and rewriting our equations as

$$\Delta \tau^2 + \Delta x^2 + \Delta y^2 + \Delta z^2 = \Delta \tau'^2 + \Delta x'^2 + \Delta y'^2 + \Delta z'^2. \quad (1.4)$$

This expression (1.4) appears to be the equality of two distances in some abstract 4D space. Let's not yet trouble ourselves with the physical interpretation of that space, but instead ponder the mathematical side of it. We are dealing with two 4D coordinate systems related by some unknown linear transformation that preserves distances between arbitrary points. There are only a handful of transformations that possess such properties: a rotation, a translation, a reflection, or any composition of the above. These transformations are the only linear operations which preserve distances between points. Can any of these transformations represent a physical transition to a moving frame?

Motion with a constant speed is clearly not a rotation nor a translation by any fixed distance, nor is it a reflection! It is also not a composition of any of these operations. So, have we overlooked anything? In order to figure this out, we need to stop chewing gum and think a little harder. Minkowski, clearly not a gum chewer, spotted another possibility we have missed: a special kind of rotation. Not just a spacial rotation, but rather a rotation within a plane spanned by the time axis τ and one of the spacial directions. This probably sounds a bit mysterious and that's ok. But, as we have run out of other options, there is nothing left but to continue! Suppose

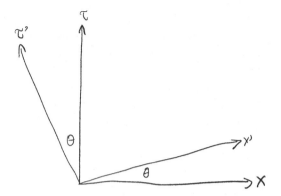

Figure 1.2:   Rotation of the coordinate system within the plane (τ, x).

that the Witch is moving relative to the traffic cop along the x axis, so let us consider a rotation within the plane spanned by (τ, x) — as shown in Fig. 1.2. If we also assume that at $t = t' = 0$ the origins of both reference frames coincided, then the rotation must take the following form:

$$\tau' = \tau \cos \theta + x \sin \theta = \frac{\tau}{\sqrt{1 + \tan^2 \theta}} + \frac{x \tan \theta}{\sqrt{1 + \tan^2 \theta}},$$

$$x' = x \cos \theta - \tau \sin \theta = \frac{x}{\sqrt{1 + \tan^2 \theta}} - \frac{\tau \tan \theta}{\sqrt{1 + \tan^2 \theta}},$$

(1.5)

where θ is an angle that somehow depends on the velocity V. All we know is that when $V = 0$ both reference frames coincide and therefore $\theta = 0$. In order to determine the full dependence $\theta(V)$ let us consider the motion of the origin of the Witch's frame: $x' = 0$. From the cop's perspective that point is moving away with the velocity V along the x axis according to the formula $x = V t$. Plugging that into the second equation (1.5) we obtain

$$x' = 0 \;\Rightarrow\; \frac{x}{\tau} = \frac{x}{ict} = \frac{V}{ic} = \tan \theta,$$

(1.6)

so the angle θ turns out to be imaginary. And there we go, we have a pseudo-rotation within a complex plane by an imaginary angle. What's the purpose of all this?! Well, let us plug the value of $\tan \theta$

into (1.5) to find out. After replacing τ with ict we get

$$t' = \frac{t - Vx/c^2}{\sqrt{1 - V^2/c^2}},$$

$$x' = \frac{x - Vt}{\sqrt{1 - V^2/c^2}},$$

(1.7)

the famous *Lorentz transformation*! Wow! The formulas are well behaved as long as we only consider frames moving with subluminal speeds: $V < c$. Otherwise some imaginary numbers pop up. The inverse transformation can be found either by manipulating the above equations or simply replacing V with $-V$:

$$t = \frac{t' + Vx'/c^2}{\sqrt{1 - V^2/c^2}},$$

$$x = \frac{x' + Vt'}{\sqrt{1 - V^2/c^2}}.$$

(1.8)

Notice that in the limit of $c \to \infty$ the equations reduce to something that is known as the *Galilean transformation*:

$$t' = t,$$

$$x' = x - Vt.$$

(1.9)

In fact, prior to the relativistic revolution nobody even wrote the equation $t' = t$. It seemed too obvious and no-one bothered to introduce a new time variable t' in the other frame. Time was understood simply as some parameter unrelated to space that was just "flowing" and the whole transformation seemed trivial. Nobody even called it a "transformation". It was "obvious" to everyone that $x' = x - Vt$. Minkowski's approach, however revealed that time actually constitutes another dimension, strongly resembling the other three spacial dimensions. The only visible difference between them was the imaginary factor i, which was introduced in order to make all the dimensions uniform. We still do not know if that is the only difference between time and space or if there are others we haven't discovered yet.

1.2 Motion as a Hyperbolic Rotation of Spacetime

All these considerations raise the following intriguing conclusion: time and space should not be treated as separate, independent entities. That's why one introduces the concept of 4D *spacetime*. Understanding this notion allows us to dip our toes in the water of peculiar developments that we have yet to face. Diverse time flow, length contraction of moving bodies, relativity of simultaneity — all these effects will become almost natural once we realise that a transition to a moving reference frame corresponds to a certain rotation of spacetime.

Questions like "what is the *real* time flow?" or "what is the *real* length?" are as meaningless as asking "what is the *real* colour of a solved Rubik's cube?" To a person looking from one side the cube looks green, to another observer it's red. So, who is right? Unfortunately, the question "who is right" is also ill-defined. Once the cube is rotated, everyone will change their mind. It is exactly the same with observations of reality from different inertial reference frames in relative motion, which corresponds to looking at spacetime at different angles.

We have previously noticed that by replacing the real time t with an imaginary time τ we can determine that the Lorentz transformation is a pseudo-rotation by an imaginary angle in a complex spacetime plane. Let us find out the geometrical interpretation of such a transformation in a "down-to-earth", real spacetime. Let us write equations (1.5) using the real time t:

$$ict' = ict \cos \theta + x \sin \theta,$$
$$x' = x \cos \theta - ict \sin \theta. \tag{1.10}$$

In order to get rid of the imaginary numbers we will use elementary identities linking trigonometric and hyperbolic functions: $\sin \theta = -i \sinh i\theta$ and $\cos \theta = \cosh i\theta$. Plugging them into (1.10) and substituting $\vartheta = i\theta$ we obtain:

$$ct' = ct \cosh \vartheta - x \sinh \vartheta,$$
$$x' = x \cosh \vartheta - ct \sinh \vartheta. \tag{1.11}$$

The geometrical interpretation of the Lorentz transformation (1.11) is given in Fig. 1.3 showing the axes of two inertial frames in a

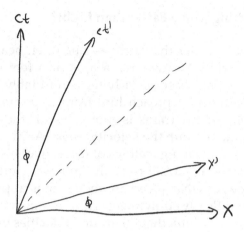

Figure 1.3: Spacetime axes of two inertial frames in relative motion.

relative motion. The orientation of these axes can be determined using the Lorentz transformation (1.11) itself. The temporal axis t' is given by the equation $x' = 0$, while the spacial axis x' is given by $ct' = 0$. From this we obtain $\tan \phi = \tanh \vartheta = V/c$, where ϕ is the relative angle of the frames' axes (as shown in Fig. 1.3). The axes corresponding to a frame approaching the speed of light, c, will approach the dashed line.

The things we've discussed so far, especially these hyperbolic rotations of spacetime, sound incredible. Why is it that a simple car journey does not provide us with any spacetime magic and the experience of moving feels pretty normal? Well, the hyperbolic angle ϑ appearing in formulas (1.11): $\vartheta = \operatorname{atanh} V/c$ can be well approximated when the considered velocity is small by $\vartheta \approx V/c$. As a consequence the transformation formulas (1.11) take the approximate form:

$$ct' \approx ct - x\vartheta,$$
$$x' \approx x - ct\vartheta. \tag{1.12}$$

From the first equation we see that the time approximately does not change $t' \approx t$, and the second equation yields $x' \approx x - Vt$ which characterises motion approximately as "changing the object's position in time". The idea of motion we are used to is just an approximated fiction.

1.3   Can Anything Move Faster than Light?

Nothing can move faster than light — one often hears. But is that really true and why? Let us have a look at a few examples. An unfortunate, suicidal observer is lying comfortably on a railway track while two trains approach him from opposite directions — see Fig. 1.4. One of the trains is moving with the velocity $0.9c$, while the other moves with the velocity $-0.9c$. As his destiny draws nearer, in addition to having feelings of regret over his predicament, the observer starts to puzzle over the following question: "from my point of view, at what speed are these trains approaching each other?" By thoughtlessly applying the rule "nothing can move faster than light", he might think that the trains' velocities must somehow add up in such a way that something smaller than c results. Well, unfortunately for this doomed soul, that would be the wrong conclusion! If we define the speed at which the trains approach each other as the rate at which their distance shrinks, then we will get $1.8c$! If, on the other hand, we ask one of the drivers to tell us the speed at which the other train approaches, he will undoubtedly respond with a velocity smaller than c (we'll return to that later).

If we really want to use the "nothing can move faster than light" rule, then the first thing we need to realise is that the word "move" really means "move relative to a given observer". In relativity a motion is usually defined with respect to a specific observer and not, as in the example above, the relative motion of two objects

Figure 1.4:   At what speed do the trains approach each other according to the suicidal observer lying on the railway track?

from the perspective of another individual. But let us consider more examples.

One hundred dwarfs arrive at Snow White's birthday party with a special gift. To celebrate the occasion, they have rehearsed a Mexican wave which they saw on TV during the 1986 World Cup. Each dwarf has a gold watch that they will use to assist with the timing of the performance. First, the dwarfs synchronise their watches and stand in a straight line, one metre apart. Then each dwarf jumps at an agreed upon time. At precisely 12 o'clock, the first dwarf jumps. Exactly one second later the second dwarf jumps, one second after that the third jumps, and so on. Now let us ask ourselves the following question: is there anything that places a limit on the maximum velocity of this wave? The answer is: not at all, the wave can spread with an arbitrary speed. To increase the velocity, the dwarfs can simply increase the distance between themselves or shorten the time delay between jumps. Such a wave could move infinitely fast if all the dwarfs jumped at once! This would not be possible if the dwarfs looked at each other and jumped only upon seeing their neighbour jump.

At least light itself should move at the speed of light when in a vacuum, right? But what if we picked up a laser pointer and aimed it at a wall? Is there a maximum velocity at which the spot of light upon the wall can move? Again, the answer is no. The laser dot itself can move arbitrarily fast without any problem. It is just a matter of how quickly we can move our hands and how far the wall is. So, what is meant by saying that "nothing can move faster than light"? We will clear things up (temporarily) in Section 1.4.

1.4 Spacetime Interval and the Causal Order

Imagine that something does move faster than light between points A and B — as shown in Fig. 1.5. That "something" could be a Mexican wave, a laser spot on a wall, a spaceship or anything else for that matter. We can see from the figure that in the unprimed frame our "something" left event A before reaching event B. However, since the trajectory of our considered object (we call such trajectories *world lines*) is oriented at an angle larger than $45°$ relative to the vertical axis, there always exist inertial frames

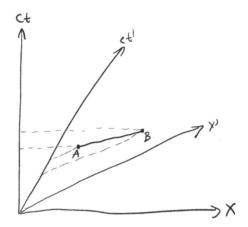

Figure 1.5:   Trajectory (worldline) of an object moving superluminally between A and B. The dashed lines represent projections of the events A and B onto the time axes of two inertial reference frames. One can see that the order of events can be reversed.

(denoted with primes in the considered figure), in which event A takes place after event B! Just look at the figure and notice how the times of the events A and B are projected on the ct' axis of the moving frame.

As a consequence, if something moves superluminally between points A and B, then some observers will see the motion from A to B, while other will witness a motion from B to A, as if the moving object went backwards in time. If the considered superluminal object was a spaceship, we would end up with a paradox: in some inertial frames the astronauts would get older as time passes, but in other frames time inside the spaceship would run backwards and the astronauts would become younger. In a drastic version of this story, in some reference frames an astronaut could even die before he is born! Since such paradoxes can be problematic, it is commonly stated that objects such as spaceships cannot move with superluminal speeds. It is also *assumed* that no signal carrying information can move superluminally, because in some inertial frames the message could be received by the recipient before it was even sent. So, the "nothing can move faster than light" rule is not really a logical necessity stemming from special relativity, but rather an additional *postulate* preventing us from having problems with causality. It is no coincidence that only inertial frames moving with speeds $V < c$ are typically considered. But we will return to that, as things become more subtle when we deal with quantum phenomena.

An alternative way of analysing the change in the order of events involves a simple application of the transformation formulas (1.7). Suppose that a signal is sent from A to B and travels the distance Δx over the time Δt in the unprimed reference frame. In the primed frame moving with the velocity V the time $\Delta t'$ elapsed between the events A and B becomes:

$$\Delta t' = \frac{\Delta t - V\Delta x/c^2}{\sqrt{1 - V^2/c^2}}. \tag{1.13}$$

We are able to add the Δ symbols, because equations (10.36) are linear. Let's take a look at the numerator of that expression: $\Delta t - V\Delta x/c^2$. That numerator is always positive for all the frame velocities $V < c$ as long as $\Delta x \leq c\Delta t$, i.e. when the signal is subluminal. But for superluminal signals, i.e. when $\Delta x > c\Delta t$, there exist frame velocities $V < c$ for which $\Delta t' < 0$, i.e. the order of events A and B gets reversed, which brings us back to the conclusions drawn from Fig. 1.5.

When we derived the Lorentz transformation, we noticed that the expression:

$$\Delta s^2 \equiv c^2\Delta t^2 - \Delta x^2 - \Delta y^2 - \Delta z^2 \tag{1.14}$$

does not change its value under the change of the reference frame, just like a length of a vector in Euclidean space does not change under rotations. Because of this invariance, the quantity Δs^2 (a *spacetime interval* "between" two events separated in time by Δt, and in space by $\Delta x, \Delta y$ and Δz) plays a very important role in relativity. In particular, a spacetime interval between emission and absorption of a light signal always equals zero. Such intervals are called *null intervals*. If the interval Δs^2 between two events is negative, these events cannot be connected by any subluminal signal, therefore they must remain causally independent (one cannot influence the other). Negative intervals are called *spacelike*. If an interval is positive — we call it *timelike*, and two events separated by it are in causal contact, since it is possible that one of them influences the other.

We have discovered that the expression (1.14) does not change its value under the change of the inertial frame, which has important consequences. Imagine that the event B is an explosion of a bomb at a given time and place. If the bomb was detonated earlier via remote control at a different location, A, then the spacetime interval between these events must either be null or timelike. Invariance of the spacetime interval leads to the conclusion that the causal sequence A–B will be preserved in every other frame, i.e. the order of these events will remain the same.

A similar rule applies to a pair of independent events separated by a spacelike interval. That interval remains spacelike in all other inertial frames, so the events are independent for all possible observers. This simple conclusion is crucial for the preservation of the causal structure of spacetime under the change of the observer.

Now is a good moment to formulate a very important principle underlying all of relativity. A principle that we have silently used without stating it explicitly. This is it: *all inertial observers are completely indistinguishable by any laws of physics.* If some event or process takes place according to one inertial observer then it must also take place for all other observers. There is no way to determine our "absolute velocity" based on any physical observations. Velocity is only a relative phenomenon. Does it sound too obvious? We will soon find out that such a simple principle has profound consequences that are far from obvious. This principle was first discovered by Galileo and it is called the *principle of relativity*.

1.5 Questions

- Is it possible to uniquely define a "rotation around an axis" in a 4D space? If so, formulate such a definition. If not, explain why.
- Are there any nonlinear transformations preserving the constancy of the speed of light?
- What is spacetime and why do we consider that notion?
- What is the geometrical interpretation of the Lorentz transformation?
- Does the result of composing two Lorentz transformations depend on the order?
- Is a composition of a Lorentz transformation for a motion with the velocity V and a transformation for a motion with the velocity $-V$ along the same direction an identity?

- How should we properly phrase the "nothing can move faster than light" rule? Does this rule follow directly from relativity?
- What role does the spacetime interval play in special relativity?
- In Fig. 1.5, why do we project the events at the temporal axis along the spacial axis and not perpendicular to the temporal axis?
- Is it possible to derive the full Lorentz transformation using just the spacial component of the coordinate transformation?

1.6   Exercises

- Write down a Lorentz transformation to the frame moving with the velocity V along the direction lying within the plane xy, and oriented at a $45°$ angle with respect to the x axis. Assume that the origins of both reference frames coincided at $t = t' = 0$.
- Consider a Galilean transformation and sketch the resulting spacetime axes of an observer moving with the velocity V along the x axis, similar to Fig. 1.3. Mark the axes' units and calculate the relative angles as functions of the velocity V.
- Verify that the following coordinate transformation:

$$t' = t$$
$$x' = \left(1 + \frac{V}{c}\right)x - Vt,$$

is linear, tends to identity for $V \to 0$ and preserves the speed of light, i.e. an object moving in the unprimed frame with the velocity c moves with the same velocity in the primed frame. Write down another transformation that has all these properties and is not a Lorentz transformation. Explain why the above transformations did not appear in our considerations as an alternative to the Lorentz transformation.
- Many years before relativity theory was discovered, Lorentz analysed properties of Maxwell's equations and noticed that the following coordinate transformation preserves the laws of electrodynamics under the change of the coordinate system:

$$t' = t - Vx/c^2,$$
$$x' = x - Vt.$$

Soon after that Poincaré corrected these equations and obtained the formulas (1.7). Investigate the properties of the transformations given by Lorentz and verify if they satisfy all the requirements formulated in the present chapter. Try to guess what motivated Poincaré to correct Lorentz's formulas.

- In our considerations leading to the derivation of the Lorentz transformation we did not discuss all the cases of possible reflections of spacetime coordinates. In 1D space a reflection is an operation with respect to a given point. In 2D space a reflection is with respect to a line. In three dimensions, with respect to a plane. In 4D space, with respect to a given volume. Consider a volume that mixes up time and space, determine the resulting reflection and propose its physical interpretation.

Chapter 2

Consequences of Time Dilation and Lorentz Contraction

2.1 Relativity of Simultaneity

Imagine holding a stick in such a way that both ends are the same distance from your nose. Your nose, your point of view, however someone else may claim that the ends of the stick are not equidistant from their nose. That is not very surprising because the stick would be oriented at a different angle relative to that observer, and a rotation of the stick would change both distances — see Fig. 2.1.

It turns out that the same principle applies to rotations of spacetime. Two simultaneous events are (in general) non-simultaneous if we start to move. Two observers in relative motion looking at spacetime at different angles should expect to obtain different outcomes of time and space measurements. For instance, if we stomp both our feet at once, an observer moving next to us may witness one of the feet stomping first. Let's check out the details. If the distance between the feet is Δx and the stomping is simultaneous, i. e. $\Delta t = 0$, then in a frame moving with velocity $V \neq 0$ along x these events are not simultaneous at all. Using the first formula in (1.7) we compute the time difference $\Delta t'$ between stomps in the moving frame:

$$\Delta t' = \frac{\Delta t - V\Delta x/c^2}{\sqrt{1 - V^2/c^2}} = \frac{-V\Delta x/c^2}{\sqrt{1 - V^2/c^2}} \neq 0. \qquad (2.1)$$

It turns out that the order of events in the moving frame depends on the direction of the velocity. The "stick" analogy allows us to

Figure 2.1: Both ends of a stick are either equidistant or non-equidistant depending on the observer. Analogously, two events that are happening simultaneously for one observer are non-simultaneous for another.

realise that the question, whether the two events are "really" simultaneous or not, is meaningless. Being simultaneous is as relative as being equidistant. There is no absolute equidistance and there is no absolute simultaneity.

2.2 Time Dilation and Lorentz Contraction

To an observer viewing a stick at an angle, the stick appears to be shorter. This is because it is rotated. Since motion is a hyperbolic rotation of spacetime, is it the same with the time lapse for a moving observer? It turns out that the time lapse in different reference frames may indeed vary. In order to prove it, let us analyse a time rate of a moving clock. Consider a Witch wearing a watch and flying on a broom at speed V. In a co-moving, primed reference frame the watch is at rest at some position x', which means that $\Delta x' = 0$. If the watch has measured a time $\Delta t'$ between two events then a resting clock in the unprimed, "resting" frame will measure a time Δt between these events that can be found with the first of the transformation formulas (1.8):

$$\Delta t = \frac{\Delta t'}{\sqrt{1 - V^2/c^2}}. \tag{2.2}$$

We find that $\Delta t > \Delta t'$. Therefore, from the perspective of a resting observer, the time flow of a moving watch seems to slow down. Although it sounds weird, the same conclusion will be drawn by the primed observer with regard to the time rate of the unprimed clock. The "moving" observer sees the "resting" clock ticking slower. Motion is relative after all (that's why we put the words "moving" and "resting" in quotation marks). We will soon return to this seemingly paradoxical conclusion and discuss it in detail.

The effect of time slowing down due to motion is called *time dilation* and it's completely analogous to the length contraction of a stick observed at an angle due to the perspective distortion. The stick appears to be shorter, because we project the space between its ends onto the direction perpendicular to the line of our sight. The time interval is shorter because we project the spacetime interval between the events onto the time axis. That's nothing other than a spacetime perspective "distortion". The effect is universal and applies the same way to any device that can be used to measure time, including a heartbeat.

A similar effect is the length contraction of moving rulers along the direction of motion. This is called *Lorentz contraction*. If we have already digested the unbelievable interpretation of motion as the hyperbolic rotation of spacetime, then the effects such as relativity of simultaneity, time dilation or Lorentz contraction should no longer surprise us. To the contrary, these effects are rather straight-forward consequences of what we already know. Quantitative study of Lorentz contraction is similar to the reasoning that lead us to the discovery of time dilation. Imagine a broom of a length $\Delta x'$ resting alongside the x' axis of the flying Witch's primed frame moving with velocity V relative to the unprimed frame. What is the broom's length Δx according to the unprimed, resting observer? We can find it using the Lorentz transformation. All we need to do is to compute the difference of broom's end positions Δx measured simultaneously ($\Delta t = 0$), using the second equation (1.7) with added Δ symbols:

$$\Delta x = \Delta x' \sqrt{1 - V^2/c^2}, \tag{2.3}$$

where V is the relative velocity of both frames. And this is the famous Lorentz contraction: the faster the broom or any other object moves, the shorter it gets.

Before we proceed, we need to clarify, why have we used the transformation (1.7) and not its inverse (1.8). Notice that if we had used (1.8) with $\Delta t' = 0$ instead of $\Delta t = 0$, we would have obtained the effect of length extension rather than contraction. To avoid such mistakes, we must realise that in order to determine the length of a moving object we should compute the difference of the positions of its ends *at the same time*. If we have positions of the endpoints of a moving broom at different times, the broom would have moved in the meantime and that would invalidate our measurements. Therefore, we need to compute (or measure) the coordinates simultaneously so that $\Delta t = 0$ and not $\Delta t' = 0$. And since simultaneity is relative, these two approaches are not equivalent.

Let us consider another fascinating feature of relativity: time dilation and Lorentz contraction cannot exist without each other. If one is real, so is the other. Here is why. Suppose our Witch is flying at a constant speed of 100 km/h between two castles separated by exactly 100 km. We might think that the trip would take the Witch exactly 1 h, but hey, there is time dilation. According to a stationary dwarf, the Witch's watch undergoes time dilation. Therefore, her watch will indicate that slightly less time elapsed during the trip. What is the Witch's explanation for an earlier arrival? After all, her own watch is not moving relative to her, so she does not witness its time dilation at all! Indeed, according to the Witch her own watch ticks at a regular rate, but the distance between the moving castles (in her frame) undergoes Lorentz contraction! The distance between the two castles gets shorter by exactly the same factor that affects the clock rate in the dwarf's frame — the so-called *Lorentz factor* $\sqrt{1 - V^2/c^2}$. And that's why both points of view: the dwarf's and the Witch's, match perfectly. Both of these observers agree that upon the arrival the Witch's watch will indicate less than an hour, but for completely different reasons. That's relativity.

Lorentz contraction always takes place *along* the direction of motion. But would it also be possible for motion to affect a perpendicular dimension? Let's assume that it does happen and then analyse the thought experiment shown in Fig. 2.2. Consider two identical, co-axial pipes approaching each other along the common axis. If such a motion affects the perpendicular dimension, for example by contracting it, then in the frame of one of the pipes the other will appear slimmer. If the relative velocity is high enough one of the pipes will pass through the other. It is clear that the question, of which pipe will go inside of which depends on the reference frame, as shown in Fig. 2.2. So, if we replace one of the pipes with a full cylinder then in one of the frames it will safely pass through the pipe, and in the other a collision will take place leading to a paradox. This simple example illustrates that the idea of a perpendicular contraction is inconsistent with the principle of relativity: various

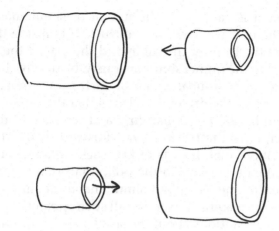

Figure 2.2: Relative motion of two pipes undergoing a hypothetical perpendicular length contraction. A view from two different reference frames.

scenarios would take place in different frames. For that reason, we need to assume that the Lorentz transformation of the perpendicular coordinates is a trivial one: $y' = y$ and $z' = z$.

2.3 Paradox of a Truck Inside a Tunnel

Lorentz contraction leads to several apparent paradoxes. Luckily all these virtual contradictions can be explained with some lightweight brain work.

The city of Warsaw is known for having a rare underground attraction — a tunnel that runs alongside the river instead of underneath it. But that's just the beginning of the paradox. Imagine a very long truck pulling a number of trailers of a total length matching the tunnel when parked inside. Next, suppose that the driver is going insanely fast — at such a relativistic speed that the policeman standing inside the tunnel concludes that the whole truck takes up only half of the tunnel's length. Violating traffic rules and speed limits is not something we encourage! However, the driver's point of view is quite different: according to him his truck does not fit inside the tunnel at all. Actually, it's the tunnel that gets contracted and half of the truck sticks outside of it. Is it not a paradox? Only an apparent one.

The problem is caused by the notion of simultaneity secretly sneaking in here. If we say that the truck "fits inside the tunnel" we mean that the front of the cabin and the rear of the last trailer are positioned in the tunnel simultaneously. Naturally, the timing of measurements of the front and back of the truck will not be simultaneous according to the driver. He claims that the front of the cabin was measured by the policeman first and the back of the trailer a moment later, after the truck moved forward. If the driver measures the position of both ends of his truck simultaneously (in his own frame) then according to the policeman standing in the tunnel, these measurements are not simultaneous at all. This example should teach us to carefully examine all such apparent "paradoxes". We must be very cautious about the sneaky notion of simultaneity.

2.4   Is Lorentz Contraction Real?

We will now consider the following important question: is Lorentz contraction real? Are all these moving objects actually becoming physically shorter? Or is it some apparent, virtual effect, perhaps some mathematical trick without any connection to reality? In order to thoroughly understand relativistic properties of reality, we have to be absolutely certain of the answer. All these things are still quite new to us, so perhaps some of us may be hesitant about the answer. The following sections will help us understand these peculiarities.

Consider the scenario depicted in Fig. 2.3. We have a stick moving horizontally along its length and a vertically moving barrier

Figure 2.3:   A stick moving horizontally through a barrier with a hole moving vertically. Positions and velocities have been chosen such that the contracted stick can pass through the hole avoiding any collision.

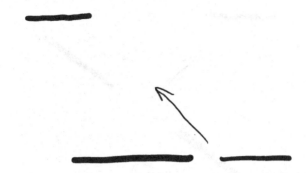

Figure 2.4:   The same situation depicted in the frame of the stick.

with a hole. Positions and velocities are chosen such that — if the Lorentz contraction takes place — collision will be avoided. But if the contraction is just some mathematical trick — the objects will collide. So, will there be a collision or not?

First of all, you might notice something puzzling here. Consider the reference frame of the stick, in which the barrier is moving at an angle. Since the barrier in this frame has a horizontal component of velocity, an additional contraction should appear along the horizontal direction — see Fig. 2.4. Therefore, we can see that, whether contraction takes place or not, the collision seems inevitable. So we should conclude that according to the principle of relativity, the collision must also take place in the first reference frame. Therefore, the contraction cannot be real.

But it turns out that we have made an interesting mistake! Figure 2.4 and the reasoning behind it are just plain wrong. The correct depiction of the situation in the second reference frame is shown in Fig. 2.5. We must remember that Lorentz contraction always takes place along the direction of motion. We cannot obtain the correct result by taking Lorentz contraction along one velocity component and composing it with the contraction along the other. Unfortunately, that's exactly what we have tried to do in the erroneous Fig. 2.4. However, Fig. 2.5 has everything correct — the Lorentz contraction takes place along the direction of velocity. And we can see that it involves a rotation of the whole barrier preventing the collision from taking place.

The reason we do not breakdown the velocity into its components and apply the transformation in two steps is quite simple.

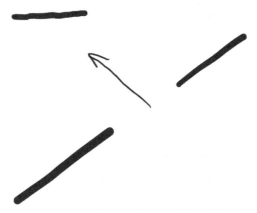

Figure 2.5:   The same situation in the rest frame of the stick — this time the correct version.

Lorentz transformation is a hyperbolic rotation and just like regular rotations it does not follow the composition law. One can conduct an interesting experiment to verify this. Just grab a book and rotate it by 90° along the vertical axis and follow that by a rotation by 90° along the horizontal axis. The result will be different to that in which the rotations were performed in the opposite order. There is a book in front of you, try it!

The inconsistency of Lorentz contraction with the principle of relativity seems apparent. But is Lorentz contraction real or not? Section 2.5 will finally clear things up once and for all.

2.5   The Paradox that Started It All

We will now consider the question that started the relativistic revolution and took physicists decades to solve. Here is how it goes.

Consider a straight electrical cable conducting an electric current. From a microscopic point of view the current is just an organised motion of negatively charged electrons. They flow with net velocity V along the cable, against the positively charged stationary ions of the metallic conductor so that the total charge cancels out. The current intensity is given by $I = e\varrho_- V S$, where e is the absolute value of the electron charge, ϱ_- is the cubic density of electrons in the conductor (equal to the density of positively charged ions, ϱ_+, since the whole cable is to be electrically neutral), and S — the cross-section area of the cable. Imagine now that at a distance r from the cable an additional electron is moving with the same velocity V along the cable — see Fig. 2.6(a). The electric current creates a magnetic field outside the cable, equal to $B = I/2\pi\varepsilon_0 c^2 r$ and the external electron will experience a magnetic force F towards the cable, equal to

$$F = eVB = \frac{eVI}{2\pi\varepsilon_0 c^2 r} = \frac{e^2 \varrho_- S V^2}{2\pi\varepsilon_0 c^2 r}. \tag{2.4}$$

This force will attract the external electron towards the conductor.

The trouble begins when we consider the same scenario in the primed inertial frame in which all the considered electrons are at rest — see Fig. 2.6(b). In this frame the electron current vanishes, but another current appears — the one coming from positive ions that are now moving in the opposite direction. That current also creates a similar magnetic field around the cable. However the problem is that the external charge is now resting, hence no magnetic force is present! How do we deal with this dilemma?

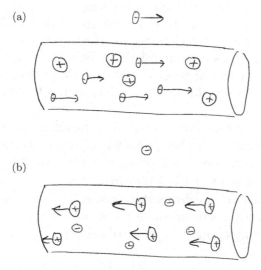

Figure 2.6:   Motion of a free electron along the cable conducting an electric current as seen from two inertial frames: (a) the rest frame of the positive ions, (b) the rest frame of the electrons. The effect of Lorentz contraction has been highly exaggerated. In reality the net velocities involved are on the order of centimetres per minute.

The problem lies deep within the formula for the magnetic Lorentz force $F = qv \times B$ acting on a charge q moving with velocity v through the magnetic field B. Let us ask ourselves: what is that velocity exactly — a velocity relative to what? After all, we can always find a frame in which that velocity vanishes. And then what, will there be no force acting on the charge?

In order to figure this out, let us go back to the unprimed frame shown in Fig. 2.6(a) and consider a piece of cable with the length L. The total charge contained in this piece of cable is composed of negative electrons with the total charge of $-eSL\varrho_-$, and positive ions with the total charge of $eSL\varrho_+$. The negative and positive charges are equal in value, but with the opposite sign; therefore, the whole piece remains electrically neutral. What happens when we start to move along the cable with the electrons? Firstly, we must observe that the electric charge of each moving electron will remain exactly the same e as when electron is at rest. If the value of the electron's charge changed with velocity, then heated bodies would immediately charge electrically. This is because some of the heat would be absorbed by the free electrons, and the rest would be transmitted to positive ions. Since the average energy absorbed by each particle is the same, the electrons would speed up much more than heavy ions. So, any dependence of the electric charge e on the particle's velocity would create an imbalance between positive and negative charges that would be easily observable given the amount of charged particles within any body. No such effect has ever been observed.

If the total electron charge (as well as the ion charge) is invariant, then could it be that the charge density changes? The length of any segment of cable containing moving electrons will be altered under the change of the reference frame from L to $L/\sqrt{1 - V^2/c^2}$ due to Lorentz contraction (electrons stop moving). Therefore, the density of electrons must decrease since the same number of electrons is now contained in a longer segment of cable:

$$\varrho'_- = \varrho_-\sqrt{1 - V^2/c^2}. \tag{2.5}$$

Similarly, the density of positive ions must increase

$$\varrho'_+ = \frac{\varrho_+}{\sqrt{1 - V^2/c^2}}. \tag{2.6}$$

The net total charge density contained within the cable in the primed reference frame equals $e\varrho' = -e\varrho'_- + e\varrho'_+ > 0$. This means that the cable is positively charged in this frame and must attract the external electron with an *electric* force! It works, at least qualitatively! In both frames, we will observe the same situation: the electron will be attracted to the cable. However, in one frame the attraction will be caused by a magnetic field, and in the other by an electric field. In Chapter 10, we will discover that relativity mixes up electric and magnetic fields and that their values depend on the choice of the observer. But the final result — their effect on the particle's motion — must be equivalent in all inertial frames.

Let's verify that this is indeed the case by quantitatively investigating the force of attraction in the primed reference frame. For this moving observer the charge density reads:

$$e\varrho' = -e\varrho'_- + e\varrho'_+ = \frac{e\varrho_- V^2/c^2}{\sqrt{1 - V^2/c^2}}. \tag{2.7}$$

The electric field generated by this charge density at the distance r is equal to

$$E = \frac{e\varrho' S}{2\pi\varepsilon_0 r}. \tag{2.8}$$

So, the electric force F' attracting the external charge towards the cable in the primed frame equals:

$$F' = \frac{e^2 \varrho_- S V^2}{2\pi\varepsilon_0 c^2 r \sqrt{1 - V^2/c^2}}. \tag{2.9}$$

We can see that both the forces: F given by (2.4) and F' given by (2.9) are nearly identical. The formulas (2.4) and (2.9) differ only by a tiny factor $\sqrt{1 - V^2/c^2}$ which, for realistic velocities, seems negligible. However, we cannot just sweep it under the rug, because that factor, however small, is still there. The situation is good, but it is not hopeless! We should remember that it is not the forces, but their effects that must be identical in both frames. In this case the momentum acquired by the accelerated electron should be the same. The correct relativistic definition of force has the form

$$F = \frac{dp}{dt},$$
(2.10)

which we will discuss in detail later. For now, we need only expect that the momentum acquired by the electron within the short time Δt (in the unprimed frame) and the corresponding time $\Delta t'$ (in the primed frame) should be identical. So, we expect the following relation to hold: $F\Delta t = F'\Delta t'$. It follows immediately from the time dilation effect (2.2) that $F = F'\sqrt{1 - V^2/c^2}$ — exactly as equations (2.4) and (2.9) tell us! And now, everything begins to match up perfectly.

2.6   The Twin Paradox

We will now discuss the famous twin paradox. It goes like this: a pair of twins, Alice and Rob, went to a space station. Rob boarded a rocket and took off on a pleasant space cruise. According to Alice, who remained at rest on the station, the time flow inside Rob's moving rocket was affected by time dilation. Therefore, Rob should have aged less than Alice during his trip. Let us designate as T the duration of time that Alice awaited Rob's return. Now, according to Rob, his entire cosmic vacation took the time τ:

$$\tau = \int_0^T \sqrt{1 - \frac{v^2(t)}{c^2}} dt < T,$$
(2.11)

known as the *proper time*, where $v(t)$ is the time-dependent velocity of the rocket. As a consequence, upon Rob's return each twin should be at a different age. Some readers might spot an apparent paradox here: from Rob's perspective it was Alice that was moving the whole time and therefore she should be younger upon his arrival. The flaw in this reasoning lies in the fact that Rob's rocket is not inertial, since it needs acceleration in order to return. Therefore, the relativistic formulas that we have used simply do not apply in this frame. Motion with constant velocity is relative, but our example shows that this is not the case anymore when acceleration is involved.

There is one way to account for Rob's point of view without diving into physics of accelerated frames (we will discuss those later). Let's consider a scenario in which Rob is moving away from Alice with a constant velocity, then accelerates back for a brief period, and then continues to move back with a constant velocity again, as shown in Fig. 2.7. Therefore, there are two stages of Rob's motion, in which we can consider him to be inertial: before he starts to accelerate in the middle of the trip, and after he has finished accelerating. In order to find out what Rob's perspective of Alice's time lapse is, we will introduce the notion of a *plane of simultaneity*, which is a set of all simultaneous events for a given observer. For instance, a resting observer's plane of simultaneity would just be a horizontal line, parallel to the x axis (if we are not taking into account the perpendicular dimensions). Any pair of events lying on that plane is simultaneous for that observer. Similarly, for any other inertial observer, a plane of simultaneity is simply any straight line parallel to his spacial axis x'. Rob's planes of simultaneity during inertial stages of his trip are shown with dashed lines in Fig. 2.7. They illustrate the fact that until the moment Rob starts to accelerate, Alice's clock is lagging behind his (in Rob's own reference frame). Soon after the acceleration is finished, Alice's clock is suddenly ahead of Rob's clock. And although from that point the time dilation makes her clock tick slower, when they meet again Alice will still be older. This study

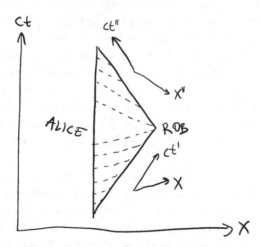

Figure 2.7: The twin paradox as seen from the Rob's perspective. His consecutive planes of simultaneity are depicted with dashed lines.

shows that, in Rob's accelerated, short non-inertial part of the trajectory, Alice's clock speeds up relative to his own clock. So, indeed the acceleration is responsible for the asymmetry between both perspectives.

There were several interesting experiments carried out in order to test the physics of the twin paradox. In one of them [2] a pair of identical atomic clocks were synchronised and then one of them was placed in a supersonic plane that flew around the world. Upon return it turned out that the moving clock had "aged" less relative to the clock that remained at rest the whole time. The results were in perfect agreement with the formula (2.11) and other predictions of relativity.

Finally, there is one more important remark to be made. We have agreed wholeheartedly with the formula (2.11) describing a time lapse on a moving clock. This formula can be understood as follows: at a given time t the rocket is moving with an instantaneous velocity $v(t)$ over an infinitesimally short time dt in which the velocity can be considered constant. Then we assume that the time dilation over that short period is the same as for a clock moving with a constant speed, yielding a time interval equal to $\sqrt{1 - v^2(t)/c^2}dt$. The formula (2.11) assumes that the rate of a moving clock only depends on its instantaneous velocity and not acceleration or higher derivatives of position. This is a very strong assumption and it is sometimes called the *clock postulate*. It must be stressed that the formula (2.11) does not logically follow from any fundamental physical laws. The time dilation effect in a uniform motion is universal and affects all clocks exactly the same way. Acceleration, on the other hand, has various effects on various clocks. A pendulum clock for instance, when accelerated too much, is likely to break. An atomic clock, which is essentially a collection of lasers and a cloud of atoms sitting on an optical table in a lab, does not like being accelerated too much either! If we accelerate it, all these fancy components will clatter to the floor.

So, what kind of clock can be completely insensitive to acceleration? Is there even such device? A hypothetical clock like this is called an *ideal clock* and the best candidate for it is an unstable elementary quantum particle, such as a muon that has no internal structure and therefore nothing within to break. By measuring the decay rate of such a particle, we can measure time. The decay rate of accelerated muons has been measured for accelerations up to $10^{18}g$ and no discrepancy with the formula (2.11) has been found [3, 4]. Unfortunately, it has been theoretically shown, that for higher accelerations this formula (2.11) must be modified — even for muons [5, 6] — due to some peculiar quantum effects. So, the notion of an "ideal clock" turns out to be just an approximation at best.

2.7 The Unbearable Loopiness of Being?

There is an old story about the inhabitants of a tiny planet, shadow-like creatures that were unfamiliar with the notion of "up" or "down". That tribe of shadows surrounded their kingdom with a wall made of shadow bricks and lived in peace. But, one day, they decided to start expanding into new territories. Every time the tribe chose to expand, they would tear down their shadow-wall and construct a new one around their newly acquired land. In order to do this, they needed to prepare extra shadow-bricks to ensure that they could extend the shadow-wall. But on one occasion, as the shadow-people set out to conquer more land, they suddenly realised that the new wall didn't need any new bricks! The shadow creatures were unable to explain this anomaly, and a rumour began to spread that the tribe had been cursed. From that moment on, the course of events only grew stranger. Each time a new wall was built, not only were no new bricks needed, but some of the old ones were left unused. There was still no explanation. The acquired terrain grew larger and larger, while the surrounding wall grew shorter and shorter. One shadow-person even calculated that if this trend continued, then within three weeks the wall would vanish completely! What would happen then? No shadow could tell. The whole planet lived in fear of that strange, impending moment. Finally, when the day had come, something unbelievable happened. All the shadow troops that set out to conquer the north met with the ones that had previously moved out towards the south, west and east. In the years that followed this strange phenomenon remained a complete mystery to the tribe of shadows. This was because they were totally unaware of the fact that their planet was a sphere (a notion they had not even invented yet).

Let us consider for a second — would it be possible for *our* space-time to be equipped with similar *periodic boundary conditions?* In such a scenario it would be possible to move straight ahead only to arrive back at the starting point after a certain time. Since light would also travel in loops, we could even see our own backs! Well, if the frame in which that happened was inertial, it would contradict the principle of relativity. And here is why: Imagine if our pair of twin inertial observers, Alice and Bob, had decided to set out on solo journeys in opposite directions at a constant speed. Which of them should be younger (if either) when they finally meet again? According to Alice, Bob was moving the whole time; so he should be younger. But in Bob's opinion it is the opposite — Alice must be younger. And we know very well that we cannot have both! What we have here is a real twin paradox, not an apparent one. The only way to resolve it is to admit that, in spacetimes with periodic boundary conditions, no principle of relativity can hold. A preferred inertial reference frame must inevitably exist in such circumstances.

Notice that, in many exotic contemporary theories, extra spacetime dimensions are often considered. The fact that we do not observe them is explained by postulating that these extra dimensions are somehow "curled up" in small circles following periodic boundary conditions. As we have just found out — these extra dimensions violate special-relativistic requirements — just so you know!

2.8   Elvis Lives!

We have already discovered that the notion of simultaneity is rather shaky. Two events are simultaneous for a given observer if they lie on a common plane of simultaneity, which is parallel to the spatial axis of the observer. Note that a moving frame of reference has its axes hyperbolically rotated, as shown in Fig. 1.3. For that reason, a pair of events lying on some plane of simultaneity of one observer does not lie on the plane of simultaneity of another. We already know all that.

For everyday motions the angle of the considered hyperbolic rotation is quite small, but for large distances even a tiny rotation can have large effect. Suppose we are walking back and forth across our room. The corresponding angle of spacetime rotation alternates between a positive and a negative value and the spatial axis x' of our moving frame tilts up and down by a small angle. But this rotation of the plane of simultaneity makes a distant star older and younger, older and younger; by nearly one minute, if the star is 100 light years away! Does it make us question how meaningful the notion of simultaneity really is? It definitely should. We are used to this notion; we apply it every day. Anytime we think about somebody else, trying to figure out what are they doing *right now* — we give some meaning to the notion of simultaneity. But it seems that it only has an approximate sense, when the distance between us and our friend is tiny. Otherwise, even the slightest motion could cause some distant object to become older or younger.

We can also ask another interesting question: is there an inertial observer, *right now*, for whom Elvis Presley is still alive? This sounds a bit weird, so let's consider this carefully. Let event A denote Elvis' death on the 17th of August 1977 in Graceland, and event B — its 44th anniversary (assumed to be today) at the same spot, as shown in Fig. 2.8. Consider an inertial observer moving away from Elvis' grave at a faraway location C (simultaneous with B according to our frame) — as shown in Fig. 2.8. The plane of simultaneity of that

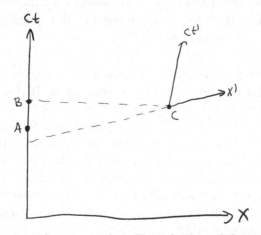

Figure 2.8: The event A corresponds to Elvis' death, and the event B is its 44th anniversary. The event C is simultaneous with B in the unprimed frame, but takes place before A in the moving frame. Planes of simultaneity of both frames are denoted with dashed lines.

observer shows that in his frame the event C takes place before A. Just look at the figure. If we take the notion of simultaneity seriously, then technically speaking, in that moving frame of reference Elvis lives. It's not magic, it's logic.

Similarly, there exists a distant inertial observer approaching us for whom our future has already happened. For such an observer, you have already finished reading this book! This is what happens if one treats the notion of simultaneity too seriously. It seems that simultaneity is not only relative, but also does not have a very deep physical meaning. In order to avoid running into such peculiar problems, we ought to use this notion with caution.

2.9   The Speed Walker Paradox

Have you ever watched a speed walking competition? These athletes walk as fast as they can without running, which can look pretty funny. But here is something weird. We know that the faster a clock moves the slower it ticks. The time dilation effect is universal, which means that not only does a speed walker's watch ticks slower, his heartbeat and hair growth is also slowed. All these "clocks" are affected by the same universal Lorentz factor. But does it mean that the faster the speed walker goes, the slower they move their feet? Is it true that when their velocity approaches the speed of light, their feet will stop moving completely?

In the reference frame of the judges, the time lapse of the speed walker's frame is affected by the Lorentz factor $\sqrt{1 - V^2/c^2}$, where V is the (instantaneous) velocity of the walk. It does not mean, however, that we are allowed to scale the *velocities* of all the elements of the moving object! Such scaling with the Lorentz factor only applies to objects that are resting in the moving frame. That's how we derived the time dilation formula (2.2) after all. The speed walkers' legs are definitely moving fast, so our derivation does not directly apply here and we should not jump to conclusions. In order to figure out what's going on here, we will introduce a simple model of the mechanics of walking.

In our problem, we are only concerned with three points of the walking body: the centre of mass (C), and the positions of the left and right foot (L and R, respectively). Let us consider the mechanics

Figure 2.9:   A toy model of walking in the co-moving reference frame.

of walking in the speed walker's own primed reference frame, in which the point C remains approximately at rest — Fig. 2.9. In this frame, the sidewalk moves backwards with the velocity $-V$ and the left foot L currently touching the ground is moving together with the sidewalk at the same velocity $-V$. At the same time, the right foot R is moving forward with the velocity V. The ground rules of speed walking specify that at least one foot has to touch the ground at any moment, otherwise the athlete will be disqualified. For that reason, a speed walker that wants to go as fast as possible must switch their feet *simultaneously*, at least in their own frame. The moment they stomp the right foot, they must also lift the left one.

Suppose that at $t' = 0$ the left foot was lifted at $x'_L = -d$ and simultaneously the right foot was put down on the ground at $x'_R = d$. The change of roles takes place at $t' = \frac{2d}{V}$ and the pattern repeats. In order to determine how the walk appears in the judges' rest frame, in which the centre of mass C is moving with a constant velocity V, we apply the Lorentz transformation (1.8) to the above coordinates.

In the first stomp the left foot is lifted at $x_L = -\dfrac{d}{\sqrt{1-V^2/c^2}}$ at the moment $t_L = \dfrac{-dV/c^2}{\sqrt{1-V^2/c^2}}$ and the right foot is put down at $x_R = \dfrac{d}{\sqrt{1-V^2/c^2}}$ at time $t_R = \dfrac{dV/c^2}{\sqrt{1-V^2/c^2}}$. Then the legs are switched and the cycle is repeated. What is interesting here is that in this frame the left foot is lifted before the right foot stomps on the ground! There is a bit of flight in that walk and the judges are perfectly within their rights to disqualify the speed walker, although he argues he did everything by the book.

According to the judges the flight time $\Delta t_\uparrow$ in each step and the distance D between the foot prints on the ground are equal to

$$\Delta t_\uparrow \equiv t_R - t_L = \frac{2d/V}{\sqrt{1 - V^2/c^2}},$$

$$D \equiv x_R - x_L = \frac{2d}{\sqrt{1 - V^2/c^2}},$$

$$(2.12)$$

and they both tend to infinity as $V \to c$. How does the walker explain the great distances he is reaching with each step? The sidewalk under his feet undergoes Lorentz contraction and the finish line gets closer, that's how! To paraphrase Neil Armstrong: a small step for a speed walker — a giant leap for the judges.

2.10   Velocity Transformation in Wonderland

Consider an object described in the unprimed frame by a trajectory $x(t)$, and in the primed frame, moving with velocity V by the transformed trajectory $x'(t')$. A relation between the corresponding velocities $v \equiv \frac{dx}{dt}$ and $v' \equiv \frac{dx'}{dt'}$ can be found by substituting the Lorentz transformation (1.7) into the definition of v':

$$v' = \frac{dx - V\,dt}{dt - V\,dx/c^2} = \frac{v - V}{1 - vV/c^2}.$$

$$(2.13)$$

Notice that the velocity v of the considered object does not have to be constant nor subluminal! After all, the object we consider could be a laser dot moving around on the wall or the dwarfs' Mexican wave. On the other hand, the velocity V of the reference frame has to be both constant and subluminal.

The obtained velocity transformation formula (2.13) is only valid when the velocity v is along the x axis. In general, when that velocity is an arbitrary vector $v = (v^x, v^y, v^z)$ then in the primed frame we will have:

$$v'^x \equiv \frac{dx'}{dt'} = \frac{v^x - V}{1 - v^x V/c^2},$$

$$v'^y \equiv \frac{dy'}{dt'} = \frac{v^y \sqrt{1 - V^2/c^2}}{1 - v^x V/c^2}, \qquad (2.14)$$

$$v'^z \equiv \frac{dz'}{dt'} = \frac{v^z \sqrt{1 - V^2/c^2}}{1 - v^x V/c^2}.$$

The obtained formulas can serve to test the constancy of the speed of light. If something (not necessarily light!) is moving in the unprimed frame with the velocity $v = c$ along the x axis, then in the primed frame moving with the traditional velocity V we have:

$$v' = \frac{c - V}{1 - cV/c^2} = c\frac{1 - V/c}{1 - V/c} = c, \qquad (2.15)$$

and everything turns out as expected!

In Fig. 2.10, we have plotted how the velocity of a moving object changes with the velocity of the observer. As we can see, the faster we chase, the slower the body escapes. For completeness we have picked the velocity in the resting frame to be $v = 0.5c$. Until now, all these things looked reasonable. It is time to discover

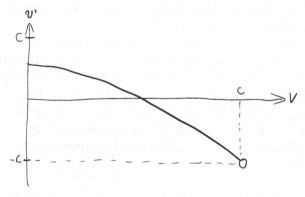

Figure 2.10: Dependence of the velocity v' of the moving object on the velocity V of the observer chasing it. In the resting frame the object moves with the velocity $v = 0.5c$. The faster the observer is chasing the object, the slower it escapes. For $V = v$ the object is at rest relative to the observer. For small velocities V the dependence $v'(V)$ is approximately linear, as in the non-relativistic mechanics.

Figure 2.11: A Witch flying above a Mexican wave made of dwarfs.

something weird! So far, we have only considered objects moving with subluminal velocities. How about checking out the eventuality of a superluminal speed? Nothing prevents us from considering the motion of a laser spot on the wall, or a Mexican wave of jumping dwarfs. Let's find out: how does the superluminal speed of this Mexican wave appear to the Witch flying on her broom with a subluminal velocity V — see Fig. 2.11.

Let the Mexican wave move with the velocity $v = 1.5c$. In Fig. 2.12, we plot the velocity v' of the wave observed by the Witch as a function of her velocity V. The consequences of the formula (2.13) are striking: the faster the Witch is flying, the faster the wave moves away! At the velocity $V = c^2/v$ the wave moves infinitely fast (all the dwarfs jump up at the same time) and, as the speed of the Witch increases, inverts the sign and becomes negative. Isn't that something?

2.11 Questions

- Consider an arbitrary pair of events. Is it always possible to find a frame of reference, in which these events are simultaneous? Is it

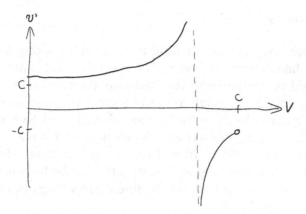

Figure 2.12: The effect of escaping dwarfs: an unintuitive dependence of the speed of a superluminal wave on the velocity of the observer.

always possible to find another reference frame, in which these two events happen at the same position?

- Consider a pair of events separated by a null spacetime interval. Does the order of these events depend on the observer?
- Is it true that for any pair of events there exists a non-trivial Lorentz transformation that does not change the time difference between these events?
- We argued that the Lorentz contraction along the perpendicular dimension is inconsistent with the principle of relativity. Can we also rule out the possibility of velocity-dependent rotation along the direction of motion?
- Are Lorentz contracted objects physically shorter, or is it a virtual effect?
- Suppose that one gives up the ideal clock postulate and looks for a more general expression for the time measured by a clock moving along an arbitrary trajectory. What would be the most general requirement concerning such an expression?
- Is it true that a transition from a resting frame A to another frame B, moving with the relative velocity V, followed by a transition from B to yet another frame C moving with the velocity V relative to B along the same direction, is equivalent to a transition from A to C moving with the velocity $2V$ relative to A?

2.12 Exercises

- A hundred rockets are hurtling through space along a common straight line one after another. The second rocket is moving with the speed $0.9c$ relative to the first one, the third one is moving with the speed $0.9c$ relative to the second one, and so on. What is the relative velocity of the hundredth rocket relative to the first one? *Hint:* the angle of hyperbolic rotation is additive.
- A spaceship travelling away from Earth with the velocity v is broadcasting a radio signal. The duration of the broadcast inside the spaceship's studio equals the time τ. How long does the transmission last on Earth?

Chapter 3

Hard Life in 3D

3.1 Lorentz Transformation in 3D

So far, we have mostly discussed problems involving just one dimension of space. Motion of the considered reference frames always took place along the "x axis". Now it is time to move on to the 3D world. The road will be bumpy and we will have to confront the *most* difficult problem in special relativity. But never fear, we will do it and we will do it together! Let's begin by recalling the Lorentz transformation in all of its glory:

$$t' = \frac{t - Vx/c^2}{\sqrt{1 - V^2/c^2}},$$

$$x' = \frac{x - Vt}{\sqrt{1 - V^2/c^2}}, \tag{3.1}$$

$$y' = y,$$

$$z' = z.$$

The above formulas define a transition from a "resting" frame to another one, moving with velocity V along the x axis — the usual stuff. But what if the velocity was along another direction? How would that affect the transformation formulas? Firstly, the only reason for x to play a special role in equations (3.1) is that the velocity is along that direction. In any other case we would have to replace x with the direction along the velocity, which is $r \cdot V/V$, where

$r = (x, y, z)$ and V is the velocity vector. This allows us to generalise the transformation formula for the temporal coordinate:

$$t' = \frac{t - r \cdot V/c^2}{\sqrt{1 - V^2/c^2}}. \tag{3.2}$$

We also know, that the spacial component of the vector r perpendicular to velocity, $r_\perp$, remains unaffected by the transformation. Only the component parallel to the velocity, $r_\parallel$, changes. This brings us to the following set of equations:

$$r'_\parallel = \frac{r_\parallel - Vt}{\sqrt{1 - V^2/c^2}} = \frac{\frac{r \cdot V}{V^2} - t}{\sqrt{1 - V^2/c^2}} V, \tag{3.3}$$

$$r'_\perp = r_\perp = r - \frac{r \cdot V}{V^2} V. \tag{3.4}$$

Since $r' = r'_\parallel + r'_\perp$, we obtain the Lorentz transformation to a frame moving with an arbitrary velocity V:

$$t' = \frac{t - \frac{r \cdot V}{c^2}}{\sqrt{1 - \frac{V^2}{c^2}}},$$

$$r' = r - \frac{r \cdot V}{V^2} V + \frac{\frac{r \cdot V}{V^2} - t}{\sqrt{1 - \frac{V^2}{c^2}}} V. \tag{3.5}$$

From now on we will be dealing with a new effect. Since the motion of the primed frame is no longer along any of the axes, the Lorentz contraction of that frame will tilt its spacial axes, so that they will *not* be mutually orthogonal anymore.

We can also derive a general form of the velocity transformation formula for the case when both the object, and the observer are moving with arbitrary velocities $v \equiv \frac{dr}{dt}$ and V, respectively. All we have to do is to compute the derivative $v' \equiv \frac{dr'}{dt'}$ using the Lorentz transformation (3.5):

$$v' = \frac{\sqrt{1 - \frac{V^2}{c^2}} \left(v - \frac{v \cdot V}{V^2} V \right) - \left(V - \frac{v \cdot V}{V^2} V \right)}{1 - \frac{v \cdot V}{c^2}}. \tag{3.6}$$

This little formula looks rather beastly and it is hard to believe that its non-relativistic limit of $c \to \infty$ simplifies to something as nice and trivial as $v' = v - V$.

Equipped with our new equations we are ready to attack.

3.2 The Hardest Problem in Special Relativity

Two inertial observers, Alice (A) and Bob (B), are watching a Witch (W) flying with constant velocity. According to Alice the Witch is moving with velocity v, according to Bob the Witch is moving with velocity v'. What is Bob's relative velocity V with respect to Alice?

That's it. Although it sounds pretty straightforward, this is the hardest problem in relativistic kinematics. Good luck trying to find a solution in any textbook. And, what makes it even more irritating, is the fact that the non-relativistic version of this problem has such a trivial solution: $V = v - v'$. So why is this problem so hard to solve? Understanding why is also a part of the problem!

If Bob is moving relative to Alice with velocity V then Alice is moving relative to Bob with velocity $-V$. We can also introduce the Witch's inertial frame, in which Alice appears to be moving with

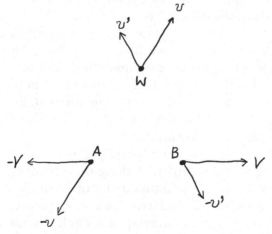

Figure 3.1: A scheme showing the relative velocities of the three reference frames: Alice's (A), Bob's (B) and the Witch's (W).

velocity $-v$ and Bob with velocity $-v'$, as shown schematically in Fig. 3.1. This allows us to apply the formula (3.6) with properly interchanged velocities. Let's consider the transformation from the Witch's frame to Alice's frame for which we need to substitute $v \rightarrow -v'$ and $V \rightarrow -v$, and the unknown velocity that we want to determine becomes $v' \rightarrow V$. Plugging these into (3.6) we obtain:

$$V = \frac{\sqrt{1 - \frac{v^2}{c^2}}\left(-v' + \frac{v' \cdot v}{v^2}v\right) + \left(v - \frac{v' \cdot v}{v^2}v\right)}{1 - \frac{v' \cdot v}{c^2}}. \tag{3.7}$$

It was not so hard, was it? At this point, some of you might stumble upon a cognitive dissonance. On the one hand every step we have taken seems legitimate, but on the other hand — should the obtained expression be an *antisymmetric* function of the velocities v and v'? Permit yourself a short break to think about this question. The answer? Yes, it *should*! And is it antisymmetric? At first glance it may be hard to say. Therefore, let us consider a special case, in which the velocities v and v' are orthogonal. Then the obtained expression (3.7) reduces to

$$V = -v'\sqrt{1 - \frac{v^2}{c^2}} + v, \tag{3.8}$$

and we can clearly see that the resulting expression is not antisymmetric at all! We must have made a mistake somewhere in our reasoning. And that mistake is quite hard to find.

We made an error when we introduced the inertial frame of the Witch. To see exactly what we have messed up, let us carefully analyse the following simple case involving three observers: Alice, Bob, and the Witch. Suppose that, according to Bob, the Witch is moving vertically up and Alice is moving horizontally to the left, as shown in Fig. 3.2. The initial positions are chosen such that all the frames' axes overlap at one instant.

And now, let's now make a transition to Alice's frame by performing a Lorentz transformation along the horizontal axis. Such a transformation changes the simultaneity of events lying along that axis, but not along the vertical axis. As a consequence, the horizontal axes of A and W will not overlap at a single instant, but the vertical axes will. This is possible if the axes of W form an obtuse angle according to Alice, as shown in Fig. 3.3(a).

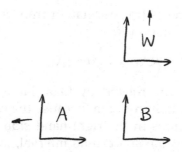

Figure 3.2: Alice and the Witch observed by Bob.

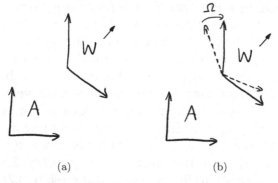

(a)   (b)

Figure 3.3: Reference frame of the Witch as observed by A: (a) the correct version, (b) an incorrect expectation denoted with the dashed line.

This leads to a problem. If W was just moving relative to A at an angle, then the Lorentz contraction should cause both the axes to spread, as shown with the dashed lines in Fig. 3.3(b). So, we have a clear discrepancy between the two versions of the story. Apparently, the reference frame of the Witch, W, is rotated by some cursed angle Ω. This angle is the source of our problems, so let us try to figure out its origin.

We have just found that the Witch's frame is not only moving relative to Alice, but it is also rotated. Therefore, it follows from our example that if we compose the Lorentz transformation from A to B, denoted by $\Lambda(V)$, where V is the relative velocity, with the Lorentz transformation from B to W, denoted by $\Lambda(v')$, then the resulting operation is the Lorentz transformation from A to W, $\Lambda(v)$, composed with an *extra* rotation $\mathcal{R}$:

$$\mathcal{R}\Lambda(v) = \Lambda(v')\Lambda(V), \tag{3.9}$$

where v, v' and V are related by (3.6). Note that $\mathcal{R}$ cannot be any other operation than rotation, because the composition of two Lorentz transformations in the right-hand side of (3.9) is a linear operation preserving the space time interval, and so must be the left–hand side of that equation. This rotation goes by the name of the *Thomas–Wigner rotation* and was discovered by accident [7] more than 20 years after the discovery of special relativity, surprising everyone, including Einstein. Such a rotation is present whenever a pair of Lorentz transformations involving non-collinear velocities is composed. We should also note that the Thomas–Wigner rotation always takes place within a plane spanned by the two composed velocities, which is the only preferred plane in our problem.

If the Witch is moving relative to Alice with velocity v then because of the Thomas–Wigner rotation Alice is not moving relative to the Witch with velocity $-v$ (or Bob is not moving relative to the Witch with velocity $-v'$). The correct velocity is rotated. That's the mistake we made in the diagram shown in Fig. 3.1. That mistake led us eventually to the erroneous expression (3.7). Before we fix that error and calculate the *correct* relative velocity, let's take a closer look at the properties of the Thomas–Wigner rotation.

3.3 Thomas Precession

The consequences of the Thomas–Wigner rotation are quite bizarre. Whenever a moving body changes the direction of velocity due to some external forces, it must rotate. This happens because the change of motion can be seen as a composition of two Lorentz transformations. Therefore, any object moving along a curvilinear trajectory has to rotate, even if no torque is directly applied. This geometrical effect is called *Thomas precession* and we will now determine its angular velocity [8, 9].

Consider Bob, B, moving relative to Alice's frame, A, with a relativistic velocity v, as shown in Fig. 3.4. If Bob's velocity changes by an infinitesimal value dv' relative to his initial reference frame B,

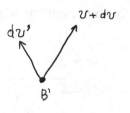

Figure 3.4: The correct scheme of the relative velocities of the three reference frames.

Alice will observe his velocity to change from v to some $v + dv$. Bob's new frame B′ is now rotated relative to Alice's frame by the Thomas–Wigner angle $d\Omega$. This is a purely relativistic effect, absent in Newtonian kinematics. In non-relativistic theory, there is also no difference between the change of Bob's velocity in A and in B: $dv' = dv$. Therefore, the Thomas–Wigner angle can be interpreted as an angle between the relativistic velocity of B′ with respect to A, $v + dv$, and its non-relativistic approximation, $v + dv'$. An angle between any two vectors can be determined by computing their vector product. For small angles we have $\sin \Omega \approx \Omega$ and we can write

$$d\Omega \approx \frac{(v + dv') \times (v + dv)}{v^2} \approx \frac{1}{v^2}(v \times dv - v \times dv'), \quad (3.10)$$

where we have used the fact that $v \times v = 0$ and neglected higher-order infinitesimal terms.

Let us use the velocity transformation formula (3.6) for the transition from A to B. We need to substitute $V \to v$, $v \to v + dv$, and $v' \to dv'$ in (3.6), which gives us

$$dv' = \frac{\sqrt{1 - \frac{v^2}{c^2}}\left(v + dv - \frac{(v+dv)\cdot v}{v^2}v\right) - \left(v - \frac{(v+dv)\cdot v}{v^2}v\right)}{1 - \frac{(v+dv)\cdot v}{c^2}}. \quad (3.11)$$

Next, let us vector multiply the result by v and use the fact that $v \times v = 0$:

$$v \times dv' = \frac{\sqrt{1 - \frac{v^2}{c^2}}(v \times dv)}{1 - \frac{(v+dv)\cdot v}{c^2}} \approx \frac{v \times dv}{\sqrt{1 - v^2/c^2}}, \qquad (3.12)$$

where we have left out higher-order infinitesimal corrections from the denominator. Substituting (3.12) into (3.10) and dividing both sides by the infinitesimal time dt we obtain the Thomas precession rate [9]:

$$\dot{\Omega} = -\frac{1}{v^2}\left(\frac{1}{\sqrt{1 - v^2}} - 1\right) v \times \dot{v}. \qquad (3.13)$$

Whenever a body moving with velocity v undergoes acceleration $\dot{v}$ that has a component *perpendicular* to the velocity, the body is subject to *rotation* with the angular velocity $\dot{\Omega}$ given by (3.13).

One of the most prominent phenomena, in which Thomas precession plays a role is the motion of an electron within an atomic orbit. In addition to its electronic charge, an electron carries an intrinsic angular momentum called *spin*. The orbital motion of the electron causes the spin to rotate with a constant angular velocity resulting from the formula (3.13). By assuming a classical model of such motion, in which the electron moves with velocity v around a circular orbit of radius R, we can determine the total angle of rotation of the spin after a full cycle using $\dot{v} = v^2/R$.

$$\Omega = \frac{2\pi R\dot{\Omega}}{v} = 2\pi\left(\frac{1}{\sqrt{1 - v^2/c^2}} - 1\right) \approx \pi\frac{v^2}{c^2}. \qquad (3.14)$$

That total angle does not depend on the radius and the final equation is particularly easy to remember, especially given its number (π).

3.4　Brute Force Approach to the Hardest Problem in Special Relativity

Let us return to the considerations from Section 3.2, where we have left an unsolved problem (actually, a problem solved *incorrectly*) of determining the relative velocity of a pair of inertial observers. We already know what our mistake was so now it is time to find the correct answer. A brute force method would involve inverting the vector expression (3.6) to determine the velocity V as a function of v and v'. It looks like mission impossible, but that is exactly our current plan. Ennobled by bitter experience, we expect V to be antisymmetric in v and v'. Let us begin by taking a scalar product of the formula (3.6) with itself. After some algebraic transformations this should lead us to the following equation:

$$\left(1 - \frac{v \cdot V}{c^2}\right)^2 = \frac{(1 - v^2/c^2)(1 - V^2/c^2)}{1 - v'^2/c^2}. \tag{3.15}$$

It shows that whenever $v < c$, we also have $v' < c$. Similarly, for $v > c$ we have $v' > c$. Assuming that we are dealing with the first case, let's determine $v \cdot V$ and substitute it back to the formula (3.6) transformed to the following form:

$$v'\left(1 - \frac{v \cdot V}{c^2}\right) - \sqrt{1 - \frac{V^2}{c^2}}\, v = V\left(-\frac{v \cdot V}{V^2}\sqrt{1 - \frac{V^2}{c^2}} - 1 + \frac{v \cdot V}{V^2}\right). \tag{3.16}$$

After several more pages of painful algebraic transformations we finally arrive at the following equation:

$$\frac{V/c}{1 + \sqrt{1 - V^2/c^2}} = \frac{\dfrac{v}{\sqrt{1-v^2/c^2}} - \dfrac{v'}{\sqrt{1-v'^2/c^2}}}{\dfrac{c}{\sqrt{1-v^2/c^2}} + \dfrac{c}{\sqrt{1-v'^2/c^2}}}. \tag{3.17}$$

The obtained compact result has some charming symmetry in it. However, we are one excruciating step away from using it to determine V. All we need to do it is to take a square of equation (3.17), compute V^2, and put it back into (3.17). With nary a word we forge through those boring calculations and finally arrive at

$$V = \frac{\sqrt{1 - v^2/c^2} + \sqrt{1 - v'^2/c^2}}{1 - \dfrac{v \cdot v'}{c^2} + \dfrac{1 - v^2 v'^2/c^4}{\sqrt{1-v^2/c^2}\sqrt{1-v'^2/c^2}}}\left(\frac{v}{\sqrt{1 - v^2/c^2}} - \frac{v'}{\sqrt{1 - v'^2/c^2}}\right). \tag{3.18}$$

Our efforts were rewarded with a nasty, but antisymmetric result, that reduces in the non-relativistic limit to: $V \approx v - v'$. Arriving here required some hardcore algebra that was only sketched above. Now we finally understand why this is the hardest problem in special relativity. In Chapter 7, we will learn how to cleverly solve it with a straightforward, two-line calculation.

3.5 Questions

- Does the Lorentz transformation only affect the components of velocities oriented along the motion of the observer? Or do the transverse components get affected as well?
- Does the direction of light propagation depend on the motion of the observer?
- Is it possible that a composition of two Lorentz transformations does not result in a Thomas–Wigner rotation?
- How may we reconcile Thomas precession with the conservation of angular momentum?

3.6 Exercises

- Determine the angle of Thomas precession after one second of motion of an electron classically orbiting a hydrogen atom in the ground state.
- Prove that the velocities v, v' and V satisfying the relation (3.6) also satisfy (3.15) and the following identities:

$$1 - \frac{V^2}{c^2} = \left(1 - \frac{v \cdot V}{c^2}\right)\left(1 + \frac{v' \cdot V}{c^2}\right),$$

$$v'^2 = \frac{(v - V)^2 - \frac{(v \times V)^2}{c^2}}{\left(1 - \frac{v \cdot V}{c^2}\right)^2}.$$

- A Mexican wave of jumping dwarfs travels along the direction $\pm s$ with an infinite velocity with respect to Snow White. That is to say that all of the dwarfs stand in a line oriented along s, and jump simultaneously. Determine the velocity V of a Witch that witnesses the Mexican wave propagating with the superluminal velocity v.

Chapter 4

Quantum Principle of Relativity

4.1 All Inertial Observers

We've spent enough time digesting the consequences of the Lorentz transformation. Now the time has come for a big surprise, as it will become apparent that we have missed something important along the way! When we first derived the Lorentz transformation (1.7), we assumed the constancy of the speed of light. As first realised by Ignatowsky [10], this assumption is *not* required in order to complete the derivation. Let's have a look. The following method of deriving the Lorentz transformation was proposed by Szymacha [11] and was further simplified and generalised in [12]. Let us first consider a $1 + 1$-dimensional case with an inertial primed frame (t', x') traditionally moving with the velocity V relative to the unprimed frame (t, x). We are looking for the most general form of the coordinate transformation between these frames that is consistent with the Galilean principle of relativity. It must be a linear transformation, so that no point in spacetime is singled out, and its coefficients must depend only on the relative velocity V. The inverse transformation involves a sign flip of the velocity V:

$$x' = A(V) x + B(V) t,$$
$$x = A(-V) x' + B(-V) t',$$

(4.1)

where $A(V)$ and $B(V)$ are unknown functions we wish to determine. The origin of the primed frame $x' = 0$, is moving according to the equation $x = Vt$. Putting that into (4.1) we obtain

$\frac{B(V)}{A(V)} = -V$, which allows us to transform (4.1) to the following set of equations:

$$x' = A(V)(x - Vt),$$

$$t' = A(V)\left(t - \frac{A(V)A(-V) - 1}{V^2 A(V)A(-V)} Vx\right). \qquad (4.2)$$

At this stage, all we can say about the unknown function $A(V)$ is that it is either a symmetric or antisymmetric function of its argument. This is because a discrete change of sign of any spacetime coordinate in the unprimed frame should result in a discrete sign change in the transformation formulas (4.2). But, since such a sign flip also affects the sign of velocity V, the quantity $A(V)$ has to be either symmetric or antisymmetric.

In order to determine $A(V)$ uniquely, consider three inertial frames (t, x), (t', x'), and (t'', x''). Let the primed frame move with the velocity V_1 relative to the unprimed frame, and let the double-primed frame move with the velocity V_2 relative to the primed one. By iterating (4.2) we obtain

$$x'' = A(V_1)A(V_2)x\left(1 + V_1 V_2 \frac{A(V_1)A(-V_1) - 1}{V_1^2 A(V_1)A(-V_1)}\right)$$
$$- A(V_1)A(V_2)(V_1 + V_2)t. \qquad (4.3)$$

Looking at the first equation in (4.2) we see that we can express the relative velocity V by calculating the ratio between the coefficient at t and the coefficient at x and reversing its sign. Applying this rule to (4.3), we obtain the relative velocity between the unprimed and the double-primed frame:

$$V = \frac{V_1 + V_2}{1 + V_1 V_2 \frac{A(V_1)A(-V_1) - 1}{V_1^2 A(V_1)A(-V_1)}}. \qquad (4.4)$$

Notice that interchanging $V_1 \leftrightarrow -V_2$ in (4.4) should lead to the expression for the velocity of the unprimed observer relative to the double-primed observer, which is $-V$:

$$-V = \frac{-V_2 - V_1}{1 + V_2 V_1 \frac{A(-V_2)A(V_2)-1}{V_2^2 A(-V_2)A(V_2)}}. \tag{4.5}$$

Comparing (4.4) with (4.5) brings us to the following identity:

$$\frac{A(V_1)A(-V_1) - 1}{V_1^2 A(V_1)A(-V_1)} = \frac{A(V_2)A(-V_2) - 1}{V_2^2 A(V_2)A(-V_2)}, \tag{4.6}$$

which must hold for any V_1 and V_2. This can only be satisfied if both sides of the equation are equal to some constant K:

$$\frac{A(V)A(-V) - 1}{V^2 A(V)A(-V)} = K, \tag{4.7}$$

which sets a constraint on possible functions $A(V)$ appearing in (4.2).

For the symmetric case $A(-V) = A(V)$, the condition (4.7) gives us $A(V) = \pm \frac{1}{\sqrt{1-KV^2}}$. Choosing the upper sign for which we get $x' \to x$ in the limit $V \to 0$, we obtain

$$x' = \frac{x - Vt}{\sqrt{1 - KV^2}},$$
$$t' = \frac{t - KVx}{\sqrt{1 - KV^2}}. \tag{4.8}$$

The new constant K remains undetermined. The case in which $K = 0$ corresponds to the Galilean universe. The case in which $K > 0$ leads to relativistic spacetime as we know it. The last case in which $K < 0$ corresponds to a Euclidean spacetime with one of the dimensions stretched by $\sqrt{|K|}$ and the derived transformation is just a regular rotation. From now on, we select $K = \frac{1}{c^2}$, which brings us to the Lorentz transformation (1.7), well-behaved for velocities $V < c$. Everything we have discussed so far in this book is just a straightforward consequence of this set of equations. However, we can see that the constancy of the speed of light turns out to be a *consequence* of the principle of relativity, and certainly *not* a necessary assumption of the theory.

There is one more mathematical possibility: the anti-symmetric case $A(-V) = -A(V)$, in which the constraint (4.7) gives

$A(V) = \pm \frac{V/|V|}{\sqrt{V^2/c^2 - 1}}$. This solution is well-behaved only for $V > c$ and leads to the following transformation:

$$x' = \pm \frac{V}{|V|} \frac{x - Vt}{\sqrt{V^2/c^2 - 1}},$$

$$t' = \pm \frac{V}{|V|} \frac{t - Vx/c^2}{\sqrt{V^2/c^2 - 1}}.$$

(4.9)

So far, we have only used the Galilean principle of relativity, which puts no restrictions on possible velocities of the observer. Solutions (4.8) and (4.9) are both linear and they preserve the constancy of the speed of light. In order to get rid of the second branch of solutions (4.9), we would need to introduce additional physical assumptions that rule them out. However, we choose not to do so. Instead, we will investigate the physical consequences of these extra solutions.

Before continuing, a few comments are in order. Firstly, let us note that both sets of equations (4.8) and (4.9) preserve the speed of light. So, any derivation of the Lorentz transformations should also lead to the possible second branch of solutions given by (4.9). If that is not the case, then either something has been overlooked or additional limiting assumptions have been applied. This also happened when we first derived the Lorentz transformation (1.7) using the Minkowski method; moreover it is a good exercise to spot that part of the derivation, in which we have excluded the second branch of solutions. Secondly, the sign in front of the equations (4.9) cannot be uniquely determined because no $V \to 0$ limit exists. The choice of sign must remain a matter of convention. Anyway, from now on we will pick the negative sign. In this scenario, equations (4.9) describe a hyperbolic rotation by the angle $\in (\frac{\pi}{4}, \frac{3\pi}{4})$, which is a continuation of the hyperbolic rotation by the angle $\in (-\frac{\pi}{4}, \frac{\pi}{4})$ corresponding to the subluminal branch of transformations. The presence of the antisymmetric term $\frac{V}{|V|}$ is vital for keeping the relativistic invariance — even though some authors make the mistake of ignoring it. The first appearance of the correct formula (4.9) in literature can be found in [13]. Thirdly, both branches of solutions form a group structure only in the considered $1 + 1$-dimensional scenario. That does not take place in the $1 + 3$-dimensional scenario [14], therefore we will carefully discuss that case separately, in a later part of this chapter. For now, let's stick to the $1 + 1$ scenario and investigate its consequences. Finally, it is worth noting that keeping only the subluminal family of observers given by (1.7) is inconsistent with the Galilean principle of relativity, which assumed that all inertial observers are equivalent, not just the subluminal ones.

It is commonly believed that considering superluminal objects or observers leads to violation of causality and results in serious paradoxes, such as the famous grandfather paradox. We will show that this is *not* the case. Although rules of causality *must* be modified, they are modified precisely in a way that is known from postulates of *quantum theory*. But before we dive into the investigation of the bizarre consequences of the superluminal branch of solutions, let's discuss what quantum theory is first.

4.2 Does the Devil Play Dice?

Imagine a beam of light falling onto a piece of plate glass. All glass transmits only some of the light and the rest is reflected or absorbed. Just have a look at your own reflection in a window! The amount of reflected, absorbed, and transmitted light depends on the particulars of the material from which the glass is made. So, let's assume that our piece of plate glass reflects exactly half of the light and transmits the rest.

To keep it simple, let's say that the light has only one colour, red for instance, and a fixed, linear polarisation. Thanks to the work of Planck, we know that such a beam of light consists of a huge number of indivisible and identical particles called *photons*. All properties of light, such as polarisation or colour, must therefore characterise each of the identical photons making up the beam of light. Tiny photons are very simple objects, and in our scenario they will be completely indistinguishable from one another: all the colour red, and all with the same polarisation. A sketch of our thought experiment is depicted in Fig. 4.1, where all the photons are beaming onto a half-transmitting glass plate to be collected afterwards by a pair of detectors.

And now let us ask a difficult question: if half of the photons are reflected, and the other half transmitted, then (if they are all the

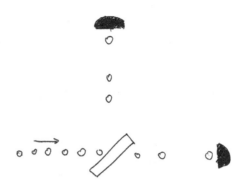

Figure 4.1:   A stream of photons falling on a half-transmitting beam splitter.

same) *what* decides the behaviour of each particular photon? Let's think about this for a moment.

The question seems rather straightforward, but it tormented Einstein until his grave. He suspected that photons had to *differ* from one another, and those differences are responsible for the fact that some of the photons bounce off the plate while others don't. Imagine, for instance, that some of the photons grow hair and the others don't. Maybe it's that hair, or some other hidden feature that decides whether a given photon gets reflected or transmitted through the glass? Suppose that's indeed the case, and our piece of glass only transmits the hairy photons and bounces off all the bald ones. Then, if we place another half-transmitting glass plate in the path of the filtered off, hairy photons, as shown in Fig. 4.2, they should all be let through! Unfortunately, numerous experiments show that this is not the case. Once again, half the photons are reflected and the other half transmitted through the glass. The results do not change one bit.

Despite countless efforts, neither this experiment nor *any* similar one was able provoke photons to change their behaviour. It was also impossible to find any particular pattern in the photon's conduct. The measurement results were no different from purely random ones, that could be obtained by flipping a coin.

So, maybe our piece of glass is not just a static object? Maybe some intrinsic dynamical processes are taking place within the glass's internal structure, causing some photons to behave one way and others differently? It turns out that such an explanation is also

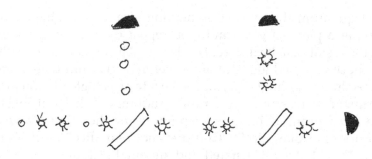

Figure 4.2:   Can single photons possess some hidden properties?

doomed to fail. Before we consider why, let's spell out what *quantum theory* has to say about the behaviour of single photons.

And it says something that'll make your hair stand on end: the reason for the particular behaviour of a single photon *does not exist at all*. A photon behaves in a way that is fundamentally unpredictable and there is *nothing* in the process of its interaction with a glass that decides its fate. The process of reflection or transmission through the glass is fundamentally random, or, as physicists like to call it, *indeterministic*.

Let's slow down a bit and digest this unbelievable claim of quantum theory. Classical physics is nothing but an effective method of predicting the future. For instance, the laws of Newtonian physics allowed one to predict the fate of a cannon ball that was shot with a given speed, at a given angle, where it's going to land, at what speed, etc. Or how soon we will see Halley's comet again after it orbits the Sun. Maxwell's theory predicted properties of light emitted by an electric charge oscillating at a given frequency. Every single law of physics was just that: a rule allowing one to predict what's going to happen in a given experiment.

This is also what our intuition is used to. Every event must have a cause. And every cause has to lead to a specific, predictable effect. If there is an A, there must be a B. If there happened to be a B, then there must have been an A that caused it.

Even if flipping a coin sometimes leads to heads and other times tails, we correctly suspect that the results are diverse only because the coin is being tossed in a slightly different way each time. And even the slightest difference at the beginning can lead to a complete change in outcome.

But quantum theory says something else. If we shine a single photon at a piece of glass and then repeat the experiment identically, the results can be different each time. The outcomes may differ from those obtained just a moment before. Quantum theory states that absolutely nothing determines the fate of a single photon in this experiment. It was this mysterious randomness that put Einstein into a tizzy; prompted him to keep muttering those famous words "God does not play dice", "God does not play dice" until his colleague Niels Bohr got irritated and snapped at him, "stop telling God what to do!"

Apparently, someone does play dice with the universe. Intrinsic randomness of the microscopical world is universal and applies to all subatomic structures. For instance, an unstable particle decays at completely random and unpredictable moments, and all we can predict in advance is the average decay time.

To expect anyone to just believe in such outrageous claims would be beyond cheeky. So, let us retreat to the seemingly reasonable hypothesis that, even if photons are identical, some processes must be taking place within the structure of the glass and these processes are responsible for the fact that the conditions of the experiment are slightly changed each time a photon falls, affecting the outcome. Perhaps it's due to thermal oscillations of the atoms — who knows?

To put this idea to test, let's conduct another simple thought experiment that will allow us to grasp the essence of all the crazy features of quantum theory.

Imagine two glass plates and two mirrors set up as depicted in Fig. 4.3. Let's place a pair of mirrors in the paths of two beams of light, which direct the beams towards a second glass plate. And just behind it, we'll place two detectors measuring the intensity of falling light. How much light will reach those two detectors if we illuminate the first glass plate with a bright beam of light?

It may appear at first that each detector will register half of the falling beam. And yet, it turns out that we have forgotten something! We can construct our experiment in such a way that all of the light will end up being recorded in only one of the detectors. But how is this possible?

Let's remember that both beams of light recombining at the second glass plate will overlap and interfere with each other. The light reaching the upper detector can do it in two possible ways.

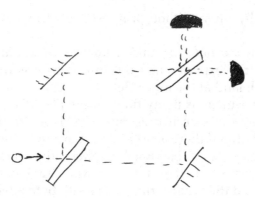

Figure 4.3:   A ray of light entering the so-called *Mach–Zehnder interferometer*.

Either it goes through an upper path, where it gets reflected three times — at the first glass plate, at the upper mirror and finally at the second glass plate. Or it can take the second, lower path that involves only a single reflection from the lower mirror. But this light has to go through two pieces of glass.

And here's the rub — light travelling through glass slows down. This means the wave travelling along the lower path will be slightly delayed compared to the wave taking the upper path. The duration of that delay depends on the thickness of the glass plates. We can arrange those plates in such a way that the resulting interference is destructive and the upper detector will remain completely dark. No light will reach it. If no light reaches the upper detector, then all of it has to end up in the lower one, because the light's energy cannot simply vanish. Let us assume that this is how we have set up our experiment, so that all of the light ends up in the lower detector.

It is time for the most interesting question: what will happen to a *single photon* if we let it fall onto the first glass plate? As a famous Polish commentator once proclaimed during a horse race, "Teraz wszystko w rękach konia!"[a]

It might seem as though the photon must somehow choose one of the paths. With no other photons to interfere with it, if we repeat the experiment many times, it will reach the upper and the lower

---

[a] "All is now in the hands of a horse!"

detector equally often, although it's still unclear what influences each choice.

It turns out we have to think again, because our conclusions disagree with the experimental outcomes; it turns out that a single photon will land at the lower detector every single time! We can repeat the experiment as many times as we like. It seems like some sort of interference is occurring, but what is interfering with our poor lonely photon if there is nothing else around? Itself?

Well, it looks like it! Having two alternative paths that could be taken, the photon behaves as if it were travelling along both paths *at once,* and then interfering with itself! Both alternative paths exhibit interference identical to that of the two beams of classical light split at the first glass plate. And as a result, our photon also ends up in the lower detector.

From what we have just observed it may follow that one photon splits into two halves, which seems to contradict our postulate that a photon cannot be divided. In order to find out whether a photon gets split or not, we can put two more detectors directly behind the first glass plate on the photon's path (or paths?). These two devices will verify which is the true path taken by the photon by making a "clicking" sound when they receive a photon. And something surprising will happen again: we will always hear only one detector click. Sometimes the upper detector will click, detecting our photon, sometimes the lower one. But never both. This whole business suggests that the photon somehow found out it was being observed and decided not to be naughty, by emerging at the end of only one path. But when nobody is looking, it behaves exactly as if it was moving along two paths at the same time. This peculiar state of the photon occurring in more than one place at once is called *quantum superposition.*

So, let's go back to the previous essential question — what determines the photon's choice of the path after the first glass plate? Is it possible that some unknown dynamical process, occurring within the glass at the moment the photon hits the plate, is responsible for photon's path selection? Some form of fluctuation, perhaps? Our conclusion that, in the absence of detectors, the photon acts as though moving along both paths at once implies the following: when the photon hits the glass plate, it makes no decision at all. Both scenarios are happening at the same time. It is only after we slide the two detectors behind the first plate that the photon decides to present itself at one place only.

So, it is the act of measurement itself that forces the photon to emerge at one side of glass plate or at another. Without measurement, the photon's position is not determined at all. If the photon's fate was decided the moment it hits the glass, there could be no interference later on! Even more disturbing is that, without exception, the moment we measure the photon's position it avoids being caught in two places at once. The unpredictability of the outcome of this measurement is, according to the doctrines of the quantum theory, a fundamental law of nature. It is not just a reflection of our ignorance about what the photon is.

More than a century has passed since the discovery of quantum theory. Over that time, not the slightest experimental deviation from that theory's predictions has been found. Although quantum theory states that it's impossible to predict outcomes of single experiments, it allows one to calculate the probabilities of particular outcomes. Currently, it is our best and most precise theory for describing nearly all reality known to us. Nevertheless, even the father and mother of the quantum theory, Bohr, used to say that anyone who is not shocked by quantum theory has not understood it. Einstein was certainly shocked, that's why he kept repeating that God does not play dice. But if it isn't God, then *who* the hell else could be doing it?

4.3 Why Does the Devil Play Dice?

Now, let's return to relativity. We have previously shown that the Galilean principle of relativity alone leads to two branches of coordinate transformations: (4.8) and (4.9) corresponding to subluminal and superluminal families of observers, respectively. In the $1 + 1$-dimensional scenario, these branches are indistinguishable. This means that a particle at rest with respect to an observer belonging to one branch will be considered superluminal by the observer belonging to the other branch — being superluminal is *relative*. We will now show that if you allow all of these observers, then a relativistic, local, and deterministic description of fundamental processes is no longer possible.

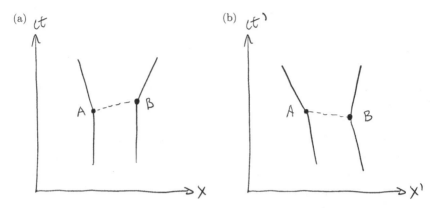

Figure 4.4:   Spacetime diagrams of a process of sending a superluminal particle as seen by two inertial observers: (a) particle emitted from A and absorbed in B, (b) the same process observed from a different inertial frame.

Suppose that a superluminal particle observed by some inertial observer was emitted from a source particle at event A and then absorbed at some later time by the identical target particle at event B — see Fig. 4.4(a). This process can also be observed from a reference frame moving with a relative subluminal velocity, depicted in Fig. 4.4(b). In this frame the superluminal particle is emitted at B and absorbed at A. Since the events A and B are separated by a spacelike interval, such inertial observers always exist.

Focus on the first frame shown in Fig. 4.4(a) and assume that the moment of emission at A can be predicted using a *local* and *deterministic* mode of description. In other words, let's assume that the past world-line of the source particle prior to event A contains locally *all* the information necessary to predict the exact moment of emission of a superluminal particle at A. Or, using Einsteinian language, there is an *element of reality* to it. On the other hand, someone next to the target particle B cannot predict the moment of absorption at B based only on local measurements of the particle B prior to the event. Particle B "does not know" it's about to be hit. Now, let us change the reference frame and study the same scenario from the perspective of the observer moving with subluminal speed, as depicted in Fig. 4.4(b). It begs the question: what *caused* the emission of the superluminal particle at the event B?

We could say that the cause of event B takes place in the distant world line of particle A. Possibly at a later time than event B itself. However, if we seek a deterministic *and* local mode of description, i.e. trying to determine the moment of emission at B only by a local measurement on particle B, it is clearly impossible. We have already assumed that the past world-line of particle B carries no information about the timing of event B. In practice, the observer with access only to the local properties of the particle B can only conclude that the emission at B was completely spontaneous and *fundamentally unpredictable*.

We previously assumed that the cause of the superluminal particle's emission at A (in the first reference frame) was determined by the past world-line of A. This assumption leads, however, to a preferred reference frame — within this frame a local, deterministic mode of description is possible; in other frames it remains impossible. It becomes extra clear when both particles A and B are *identical*, in which case we should have a perfect symmetry between the view of both inertial observers shown in Figs. 4.4(a) and 4.4(b). To preserve the Galilean principle of relativity, we must abandon our assumption that emission at A in the first frame could be determined by a local process. Consequently, we conclude that *no relativistic, local and deterministic description of the emission of a superluminal particle is possible*. If such an emission was to take place, it would appear completely random to any inertial observer. Even if we had a source of superluminal particles at our disposal, we could *not* use it to send any information, as we'd be unable to *control* the emission rate using any local operations. Therefore, although superluminal particles interacting with regular matter are considered, no obvious grandfather paradox arises.

Non-deterministic behaviour is not only a property of superluminal particles; subluminal particles also display non-deterministic behaviour. Consider the decay of one such particle into a pair of other subluminal particles, as depicted in Fig. 4.5(a). Let us picture the same process as seen by the infinitely fast moving inertial observer, for whom the transformation (4.9) reduces to

$$x' = ct,$$
$$ct' = x.$$

(4.10)

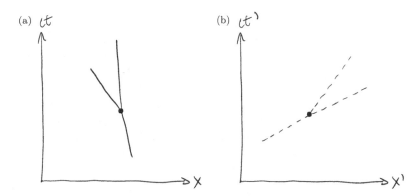

Figure 4.5:   A spacetime diagram of the decay of a subluminal particle into a pair of subluminal particles: (a) in a subluminal reference frame, (b) in a superluminal reference frame.

For such a frame, the considered decay process is depicted in Fig. 4.5(b). From this perspective all particles are superluminal, hence the decay cannot be described using any local and deterministic theory. By invoking the Galilean principle of relativity, we realise that the same must hold true for any subluminal reference frame in which all involved particles are also subluminal.

The full mathematical structure of the Lorentz transformation contains both subluminal (4.8) and superluminal (4.9) terms. The superluminal part is usually discarded, on the premise that it makes no physical sense. By ignoring superluminal observers we can obtain the familiar classical picture of a particle moving along a well-defined path. Here we have demonstrated that, if we retain the superluminal terms and take the resulting mathematics of the Lorentz transformation seriously, then we must abandon the notion of a fully deterministic reality. Next, we will also show that a scenario involving particles always moving along single paths may no longer be valid. Let's see why scenarios in which particles move along several world lines *at once* are inevitable.

4.4   Superposition of World Lines

Now let us show that if the Galilean principle of relativity involving both families of inertial observers (4.8) and (4.9) is assumed, then the emergence of superpositions of world lines is inevitable.

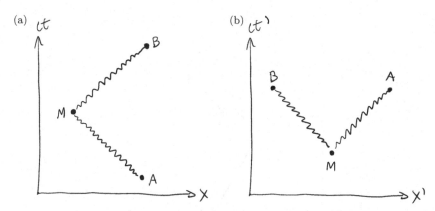

Figure 4.6: A spacetime diagram of a photon (wavy line) being reflected from a mirror: (a) in a subluminal reference frame, (b) in a superluminal reference frame.

Consider a photon emitted from a source at A, reflected from a mirror M and then received at B, as shown in Fig. 4.6(a). Suppose we place detectors in the photon's path. If a detector placed in path A–M absorbs the photon, then a similar detector placed in path M–B will not register anything, because the photon has already been absorbed. Similarly, if a detector at M–B absorbed the photon, then certainly the photon could not also have been detected at A–M. Now let us analyse the same scenario from an infinitely fast moving reference frame by applying equations (4.10) — see Fig. 4.6(b). In this reference frame the photon is travelling from M towards A and B along two paths, but if we try to detect it with a pair of detectors placed at M–A and M–B then only one of them will absorb the photon. However, as long as we do not make any observation, the motion of the photon is characterised by *two* simultaneous paths — not just one!

Apparently, even if we start with the idea of a classical particle moving along a single path, all it takes is changing of the reference frame to end up with a scenario involving more than one path. Consider the process depicted in Fig. 4.7 in which a particle emitted at A is scattered in α, where it starts to follow two paths at once towards B and B'. Viewed from the infinitely fast-moving frame, the same process will involve the particle following three paths at once. Iterating this concept will lead to scenarios involving additional simultaneous paths. Once both branches of transformations

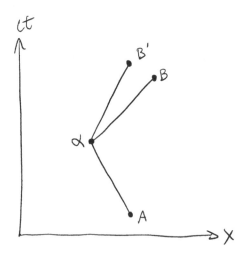

Figure 4.7: A particle emitted at A is scattered at α into motion along two paths, towards B and B' simultaneously.

(4.8) and (4.9) are involved, the classical description of a particle that *always* moves along a single trajectory becomes inconsistent with the Galilean principle of relativity.

4.5   1 + 3-Dimensional Case

The situation becomes even more interesting in the 1 + 3-dimensional case. It was shown that the smallest group involving both subluminal and superluminal 4D transformations is $SL(4, \mathbb{R})$ [14]. This cannot be a *symmetry group*, because it involves transformations which are not symmetries, such as direction-dependent time dilation. Therefore, superluminal transformations in 1 + 3-dimensional spacetime will not be symmetries. According to one interpretation [15], unlike the 1 + 1-dimensional case, the family of superluminal observers can be distinguished from the subluminal observers and therefore being superluminal is *not* a relative notion anymore. The spacetime interval in the frame (ct', x', y', z') moving superluminally:

$$c^2dt^2 - dx^2 - dy^2 - dz^2 = -c^2dt'^2 + dx'^2 - dy'^2 - dz'^2 \quad (4.11)$$

has a non-Euclidean spatial component $dx'^2 - dy'^2 - dz'^2$, which can be physically distinguished from the Euclidean space of the subluminal observers. This creates a physical difference between subluminal and superluminal observers.

Since the Galilean principle of relativity stating that all inertial frames are equivalent does not hold true in $1 + 3$-dimensional spacetime, let's consider a "quantum" version of the principle of relativity. We postulate that *the existence or non-existence of a local and deterministic mode of description of any process does not depend on the choice of inertial reference frame*. For example, let's look back at Fig. 4.4(b); if in one frame there is no local deterministic mechanism (or "element of reality") behind the particle decay in Fig. 4.4(b) in the past world-line of B in one frame, there should be no such mechanism in any other frame. This way all the conclusions of the previous sections are still valid, while still allowing for the two families of observers to be physically distinguishable.

Lastly, we would like to consider a different interpretation of the relationship between spacetime intervals in subluminal and superluminal reference frames, given by (4.11). Note that the signs of individual terms on the right-hand side of equation (4.11) suggest that the temporal coordinate cdt' will have the same properties as dy' and dz'. The quantity cdt' can be identified as a temporal coordinate, because its axis ct' must coincide with the world line of the superluminal observer. This suggests that the remaining coordinates, y' and z', are also temporal, and that there is only a single spatial dimension in a superluminal frame of reference, x'. Within such an interpretation, the interval in the $n + m$-dimensional spacetime, defined as: $ds^2 \equiv c^2 \sum_{i=1}^{n} dt_i^2 - \sum_{i=1}^{m} dr_i^2$ changes its sign for the superluminal coordinate transformation, and the perpendicular spatial coordinates change their character, thus transforming the $n + m$-dimensional spacetime into the $m + n$-dimensional one.

This disturbing property of superluminal observers doesn't just explain why they are physically different from subluminal reference frames; it also offers an interesting insight into the origin of wave properties of matter, providing a novel interpretation of the *Huygens principle*. According to that principle, *any* point at which a particle wave arrives becomes the origin of a new spherical "matter wave".

Since all known matter (and light) follows this principle, it seems as though all physical objects are compelled to travel in all directions of space from any point they visit. But when observing this peculiar behaviour from a superluminal reference frame, it looks more like all objects are forced to move symmetrically in all three "directions of time" which in some way sounds more suitable.

The $1 + 3$-dimensional Lorentz transformation between two subluminal observers is given by (3.5). The inverse transformation is obtained by substituting $V \rightarrow -V$, as well as $r \leftrightarrow r'$ and $ct \leftrightarrow ct'$.

A similar generalisation can be carried out for superluminal transformations (4.9). By replacing Vx in (4.9) with $V \cdot r$, we obtain the general transformation between a subluminal reference frame (ct, r) and a superluminal one (ct', x') moving with a superluminal velocity V:

$$
\begin{aligned}
x' &= \frac{Vt - \frac{V \cdot r}{V}}{\sqrt{V^2/c^2 - 1}}, \\
ct' &= r - \frac{V \cdot r}{V^2}V + \frac{\frac{V \cdot r}{Vc} - \frac{ct}{V}}{\sqrt{V^2/c^2 - 1}}V.
\end{aligned}
\tag{4.12}
$$

The inverse transformation to (4.12) is obtained by reversing the above set of linear equations, which is equivalent to substituting $V \rightarrow -V$, as well as $r \leftrightarrow ct'$ and $ct \leftrightarrow x'$. The equations (4.12) transform the spacetime interval according to the following equation:

$$
c^2 dt^2 - dr \cdot dr = dx'^2 - c^2 dt' \cdot dt'. \tag{4.13}
$$

For the infinite speed limit $V \rightarrow \infty$, the above formulas reduce to

$$
\begin{aligned}
x' &= ct, \\
ct' &= r,
\end{aligned}
\tag{4.14}
$$

regardless of the direction of the infinite velocity V.

Ruling out the superluminal family of observers from special relativity, regardless of whether such observers physically exist or not, is not necessary; it leads to a classical description of a particle moving along a well-defined single trajectory. However, this is in *disagreement* with experimental observations. In contrast, if one keeps both subluminal and superluminal solutions, then non-deterministic behaviour and non-classical motion of particles arise as a natural consequence.

There are plenty of other interesting things one can derive from the extended principle of relativity [12]. Among them is the correct expression for quantum probability amplitudes, along with the fact that those amplitudes must be complex numbers. But discussing these matters would take us rather far afield, so let's leave that conversation for another occasion.

4.6 Questions

- What postulates are needed to derive the Lorentz transformation?
- What does it mean for a physical system to be in a quantum superposition?
- What does it mean that "nothing can move faster than light"?
- Is it possible to send superlumimal signals using superluminal particles?

4.7 Exercises

- Solve "the hardest problem in special relativity" from Section 3.2 for the case of a superluminal Witch.

Chapter 5

Hard Bodies

5.1 Every Stick Has Two Ends (But a Slingshot Has Three)

Imagine a very long, light, and rigid stick. If we hold one end of the stick and swing it around, then the far end of the stick can move very quickly. Even a small twist of the wrist will suffice. If the stick is sufficiently light, the applied force does not have to be strong. But would it be possible to move the far end with an arbitrarily high velocity? The answer is no. And it is not because of any practical limitation. The problem has a more fundamental nature. When we initiate motion on the end of the stick we are holding, the far end will not start moving at once. First, a "sound" wave has to travel along the stick in order to put consecutive parts of the stick into motion. Therefore, the far end will not instantly "find out" that the movement has been initiated; it will stay at rest for a while, until the wave reaches it. As a consequence, the stick will bend — no matter how stiff it was meant to be. The exact motion depends on the details of the material and may be quite hard to determine. But all we need to know right now is that the notion of "rigidness" cannot be married with relativity. Ideal "rigid" bodies by definition have to react immediately to an external force, with all their volume at once. Such an instantaneous reaction is in clear contradiction with the assumption that all signals must propagate with subluminal speeds.

5.2 A Pole Vaulter Runs into a Barn

Let us go back to the Warsaw tunnel paradox from Section 2.3. In the rest frame of the truck, the tunnel is contracted — if the speed is sufficiently high, the tunnel becomes too short for the truck to fit into. From the point of view of the policeman standing inside the tunnel, the truck is contracted; beyond a certain speed it will be able to fit completely inside the tunnel. To solve the apparent paradox, we noticed that simultaneity had a different meaning in these two frames. Nevertheless, a careful reader will spot another interesting problem.

Consider a scenario in which a running pole vaulter holds his pole horizontally. He is running into an open barn of the same rest length as the rest length of his pole — see Fig. 5.1. The athlete intends to fit the pole inside the barn by way of to Lorentz contraction, but in his own frame of reference it is the moving barn that is contracting! The task seems to be impossible, as shown in Fig. 5.1(a). The same scenario, in the rest frame of the barn, seems to lead to an

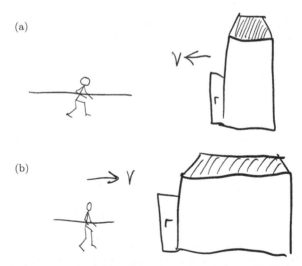

Figure 5.1:   A running pole vaulter from the point of view of (a) his own reference frame, and (b) the barn's rest frame. In the first case, the barn undergoes Lorentz contraction and the pole has little chance of fitting inside. In the barn's frame, however, the pole is contracted and will easily fit inside the barn so that the door can be shut.

opposite conclusion. In this frame, depicted in Fig. 5.1(b) the pole is contracted and will fit easily inside the barn. We can even shut the door once the vaulter and his pole are inside. It appears that we have a stark discrepancy between these two points of view, which seems to contradict the principle of relativity. So, how do we resolve this?

The situation in the rest frame of the barn seems rather straight-forward. The pole vaulter will easily fit inside, and it will be possible to close the door behind him. Once he hits the rear wall, things start to get messy. What matters is that, according to the principle of relativity, the same scenario must take place in the rest frame of the pole — and this is indeed the case. We must remember that, at the moment the front end of the pole hits the wall, the rear end does not immediately "know" about it — it remains at rest for a while longer. By the time information about the collision reaches the rear end of the pole, the door of the moving barn will already have passed it by and shut, locking the pole inside. To prove it, suppose that the rest length of the pole, P, and the rest length of the barn, B, are such that when the pole is moving with velocity V, it becomes shorter than the barn:

$$P\sqrt{1 - V^2/c^2} < B. \tag{5.1}$$

Let us move on to the pole's rest frame and assume that the signal about the collision travels as fast as possible — with the speed of light, c. The time it takes for the signal to reach the other end of the pole in its rest frame is simply: $t_{\text{sig}} = P/c$. On the other hand, the time it takes for the open door to pass the rear end of the pole is given by: $t_{\text{door}} = (P - B\sqrt{1 - V^2/c^2})/V$. To avoid conflict with the principle of relativity, the condition: $t_{\text{sig}} - t_{\text{door}} > 0$ should be satisfied:

$$
\begin{aligned}
t_{\text{sig}} - t_{\text{door}} &= \frac{P}{c} - \frac{P - B\sqrt{1 - V^2/c^2}}{V} \\
&= \frac{B}{V}\sqrt{1 - V^2/c^2} - \frac{P}{V}\sqrt{1 - V^2/c^2}\sqrt{\frac{1 - V/c}{1 + V/c}} \\
&> \frac{B}{V}\left(\sqrt{1 - V^2/c^2} - \sqrt{\frac{1 - V/c}{1 + V/c}}\right) > 0,
\end{aligned}
\tag{5.2}
$$

where we have used inequality (5.1). As we can see, the rear end of the pole will not have enough time to react to the collision of the front end before the door is shut. Paradox solved.

5.3 Two Squares Paradox

There are plenty of interesting paradoxes related to the alleged rigidity of the "rigid body". Here is another sample. Consider two squares moving within their common plane. Suppose that in the rest frame of the left square, the right one is moving along its diameter, as shown in Fig. 5.2(a) and by Lorentz contracting it becomes a rhombus.

According to the figure, the squares will inevitably collide and the first contact will take place between a vertex of the left square and the edge of the right square, leaving a clear mark of a collision. The same situation depicted in the rest frame of the right square is shown in Fig. 5.2(b). It is clear that the first impact will take place between the right vertex and the left edge, leaving the mark of the collision elsewhere. Since the location of the dent cannot depend on the choice of the reference frame, there must be a flaw in our reasoning somewhere. So, what is the correct description and where did we make the mistake?

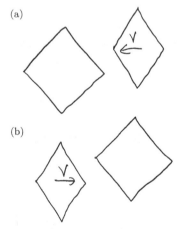

Figure 5.2: Collision between two squares as witnessed within their respective rest frames.

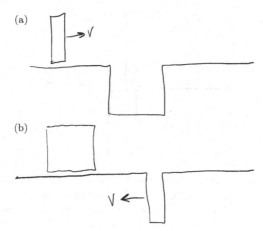

Figure 5.3: A block is moving along the table towards a whole in two alternative reference frames: (a) table's rest frame, (b) block's rest frame.

In the rest frame of the centre of weight, both squares are moving symmetrically with opposite velocities and undergoing identical Lorentz contractions, which turns them into identical rhombuses colliding at their edges. In this frame the earlier-mentioned vertices will collide simultaneously. Therefore, in any other frame the instantaneous point of collision must move along the edge with a speed higher than c. The consecutive colliding pieces will not have time to "find out" that their neighbours have already collided, so no kinetic energy will be dispersed and the vertices will not leave any clear dents.

5.4 Block on a Table with a Hole

Here is another interesting paradox involving relativistic motion of a "rigid body" [16]. Imagine a square block moving towards a hole on a table — see Fig. 5.3. In the rest frame of the table the block contracts and will easily fit inside the hole as it falls into it — see Fig. 5.3(a). On the other hand, in the rest frame of the block it is the hole that Lorentz contracts, making it seem impossible for the block to fall in — see Fig. 5.3(b).

This problem turns out to be difficult to solve in detail, because once the block reaches the hole it will experience a

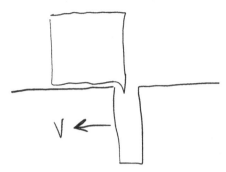

Figure 5.4:   A piece of the block above the hole starts to slip into it although the rest of the "square" is still on the table. The reaction of the material, or its "stiffness", kicks in only after a period of delay. This complicated process involving internal forces within the "square" block creates motion that is hard to determine.

torque-generating, complicated motion down into it. In its own rest frame, however, the block is too long to fit in its entirety above the moving, contracted hole. Does that mean that in this frame it's impossible for the block to fall inside the hole?

We would have a paradox here if the block was an actual rigid body. Fortunately that is not possible, and once the block reaches the the hole it will start to "pour in" despite the fact that the remaining part of the block is still on the table — see Fig. 5.4.

Internal forces within the block will eventually react, leading to complex dynamics in its body, but it will be too late to prevent the block from falling into the hole. So, although there is no obvious paradox going on here, it is still difficult to predict the exact dynamics of the block.

5.5   Internal Reaction Forces

We have already learned the motion of an extended body can initiate deformations that generate internal reaction forces resulting in further deformations. Such internal forces are only absent when the given body moves with a constant speed, despite Lorentz contraction. The reason is very simple: in the rest frame of the body there are no internal forces, so they cannot appear just because an observer started to move.

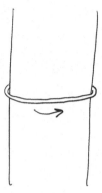

Figure 5.5:   A thin circle rotating with a relativistic speed around a stationary core.

But internal forces can also appear in the complete absence of deformations. Imagine the following, peculiar case involving a thin circle rotating around a stationary core, as we show in Fig. 5.5. By analysing each piece of the circle we could conclude that, due to its motion, it should Lorentz contract. In such a scenario, we would expect the circumference of the circle to shrink through a decrease in diameter. However, the core constrains that from happening. The inability of the circle to Lorentz contract will generate internal reaction forces that can break the material.

5.6   Questions

- Do bodies that move with a constant speed undergo internal reaction forces due to Lorentz contraction? How about bodies that rotate with a constant angular velocity?
- Consider an instantaneous force applied to a single point of an extended body. Is it true that, for a very short time, the body will react in a similar way to a fluid?

5.7   Exercises

- Consider Section 5.3's "colliding squares" paradox from the rest frame of one of the squares. Calculate the speed at which the colliding squares' contact point moves along the edge.

Chapter 6

Optical Illusions

6.1 Doppler Effect

Appearances can be deceiving. We have learned that time and space are relative. Simultaneity is relative. But wait, there's more! Usually we experience the world around us through light emitted or reflected by surrounding objects, not through measuring directly. So, let's investigate how relativity affects that light. We know that the speed of light does not depend on the velocity of the source or observer, but the light's wavelength does. Hence the famous red shift of light emitted by stars and galaxies moving away from us, a phenomenon called the *Doppler effect*. The Doppler effect may even explain why the lights of approaching cars are white, while those on cars moving away are red.

Consider a resting source of light of a wavelength λ_0. Our first objective is to determine the wavelength λ of that light observed in another reference frame, in which the source is moving with velocity v. Light is just an electromagnetic wave. Imagine that two consecutive maxima of that wave were emitted in that frame at the instants t_A^e and t_B^e while the moving source was at the distances r_A and r_B from the inertial observer, respectively. The moments at which these two consecutive maxima of the electromagnetic wave reached the observer were: $t_A^o = t_A^e + \frac{r_A}{c}$ and $t_B^o = t_B^e + \frac{r_B}{c}$. Therefore, the registered wavelength, λ, equalled:

$$\lambda = c(t_B^o - t_A^o) = c(t_B^e - t_A^e)\left(1 + \frac{r_B - r_A}{c(t_B^e - t_A^e)}\right) = \frac{c(t_B'^e - t_A'^e)}{\sqrt{1 - v^2/c^2}}\left(1 + \frac{v_r}{c}\right)$$

$$= \lambda_0 \frac{1 + \frac{v_r}{c}}{\sqrt{1 - v^2/c^2}}, \tag{6.1}$$

where v_r is the radial component of the source velocity, and the primed coordinates correspond to the inertial frame co-moving with the source. The obtained formula differs from its non-relativistic counterpart by the Lorentz factor $\sqrt{1 - v^2/c^2}$. Presence of that factor leads to some interesting consequences. For example, when a light source orbits an observer at a fixed distance, the Doppler effect causes its spectrum to red shift. This can be seen by substituting $v_r = 0$ into (6.1). Electromagnetic waves emitted by sources moving away from that observer will also red shift, whereas those emitted by sources approaching the observer are usually blue shifted. However, a source approaching the observer along a carefully chosen spiral may also be red shifted, or not Doppler shifted at all. It is all a matter of proportions between components of the velocity.

6.2   What a Drag — Light in a Moving Medium

Another surprise is ahead of us. Whenever one says that the speed of light is "always c", one has to add the sacramental "in a vacuum". It is known that, in transparent mediums such as water, light moves slower than c — and it is absolutely possible that, within such a medium, some objects move faster than light. The speed of light is given by c/n, where $n \geq 1$ is a coefficient characterising the medium. That is, for the medium at rest. And what about the speed of light within a moving medium? Is it still c/n? Not at all.

In order to obtain the correct value of the speed of light in a moving medium, we need merely transform it from the medium's rest frame. Consider light propagating in a resting medium with velocity c/n along the x axis. The velocity in a frame, in which the medium is moving with V along x can be obtained by applying the formula (2.13):

$$v' = \frac{c/n + V}{1 + V/nc} = c\frac{c + nV}{nc + V}. \tag{6.2}$$

If the medium is moving against the light with velocity $V = -c/n$, then in this frame the light will simply stay at rest! Who would have thought? The effect looks exactly as if the medium is dragging the light propagating within, much like a flowing river drags a swimming fish.

6.3   A Circle in the Shape of a Sausage

One could perhaps imagine that a Witch flying on a relativistic broomstick will appear shorter due to Lorentz contraction [17]. Nothing could be further from the truth [18–20]! The Witch will be shorter indeed, but what we will actually see is a completely different story. Sound non-sensical? The shape of a moving body and its appearance differ because the light emitted from a moving source does not reach the observer in a single instant. Let's investigate the details.

We begin by studying the appearance of a circle moving along its diameter. It has been established that the shape of a moving "circle" is an ellipse. Our objective is to determine what a snapshot of such an object would look like. In order to proceed, we first invoke the equations of the circle in its rest frame: $x'^2 + y'^2 = R^2$ and $z' = d$, where R is the radius and d is its distance from the xy plane. By applying the Lorentz transformation (1.7), we determine the set of equations characterising the shape of the circle in a frame in which it is moving with the velocity v along x:

$$\frac{(x - vt)^2}{1 - v^2/c^2} + y^2 = R^2, \quad z = d. \tag{6.3}$$

As expected, these equations describe an ellipse moving within the plane $z = d$.

Let us now consider light rays emitted by such an ellipse at (ct, x, y, z) and reaching a camera at the origin of the coordinate system, at the instant t_0. These rays are described by the equation: $x^2 + y^2 + z^2 = c^2(t_o - t)^2$. If we treat the resulting equations as a set, then its solution will be a collection of points in space, from which the light rays reaching the camera were emitted. In order to find these points we extract t from the last equation and substitute it into (6.3), to obtain the following fourth-order equation for x and y:

$$\frac{\left(x - vt_o + \frac{v}{c}\sqrt{x^2 + y^2 + d^2}\right)^2}{1 - v^2/c^2} + y^2 = R^2, \tag{6.4}$$

parametrised by t_o. The resulting curve can be found numerically — see Fig. 6.1, in which we plot the set of points from which the emitted light reached the camera at various times t_o. We have chosen the radius $R = 1$, the velocity $v = 0.9\,c$, and the distance between the plane of motion and the camera to be $d = 0.5$. From our plots it follows that the circle approaching the camera will appear to be *stretched* not contracted and that, after passing the camera, the circle will take the shape of a sausage.

The actual photographs of the moving circle may differ slightly from the plots in the Fig. 6.1 if the camera is pointing towards the moving object and not facing the circle's plane of motion. On top of that, the colours of the circles will be affected by the Doppler effect. The parts of the circle approaching the camera will be blue shifted, while those moving away will be red shifted leading to a rainbow effect.

Let us now conduct a similar study for an arbitrary shape characterised in its rest frame by a generic equation $F(x, y) = 0$, and moving within its own plane with an arbitrary velocity v along x. For such an object the illusory shape registered by the camera at t_o can be determined by following the exact same steps as before:

$$F\left(\frac{x - \frac{v}{c}\left(ct_o - \sqrt{x^2 + y^2 + d^2}\right)}{\sqrt{1 - v^2/c^2}}, y\right) = 0. \tag{6.5}$$

To make things more fun, let's have the shape be a moving bicycle [18] — the one showed in Fig. 6.2. Since the bicycle is composed out of a bunch of circles and straight lines described by elementary equations, the analysis need not be too complicated. We

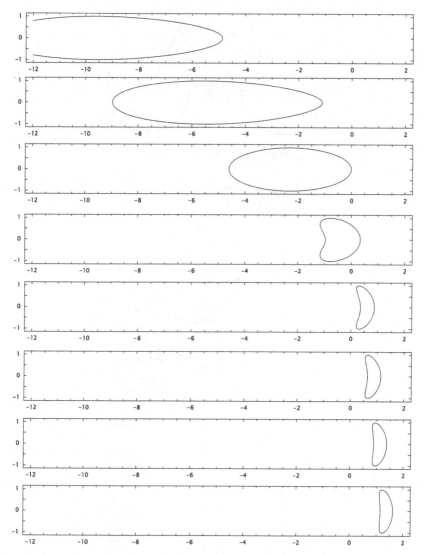

Figure 6.1: Snapshots of a moving circle. The photographs are taken at equal intervals of time, with the circle at velocity $0.9\,c$, and the plane of motion at the distance $0.5\,R$ from the camera.

use the formula (6.5) on each element separately and then put the resulting images together — see Fig. 6.3 showing a timelapse of the bicycle riding at the speed $v = 0.8\,c$ at a distance equal to the diameter of the wheel. The resulting photographs are quite bizarre.

Figure 6.2:   A bicycle at rest.

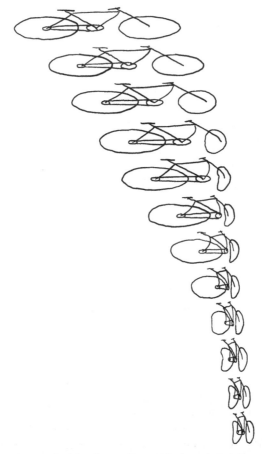

Figure 6.3:   Timelapse of a bicycle moving with the relativistic speed of $v = 0.8\,c$, taken from the distance equal to the diameter of the wheel, at equal intervals of time.

The third image from the top shows the bicycle at the closest distance from the photographer. The earlier images present the bicycle approaching the camera, and the latter ones — the bicycle moving away.

6.4   Sphere of a Shape of a Sphere

It turns out that the appearance of a moving sphere is even more interesting. It takes a shape that you'd never expect: a *sphere*! We are about to prove that the outline of the moving sphere is always circular, regardless of its velocity or position.

Consider a resting sphere at an arbitrary position in a primed frame [19] — see Fig. 6.4. Using an equation characterising the light rays tangential to the sphere that reach the observer at the origin of the coordinate system, we can express the circular outline of the sphere — as depicted in Fig. 6.4. If the unit vector pointing away from the observer towards the centre of the sphere is a', the tangent vector pointing from the observer towards the outline of the sphere is $r' = (x', y', z')$, and the angle between a' and r' is θ', as shown in Fig. 6.4, then the equation of the light ray observed

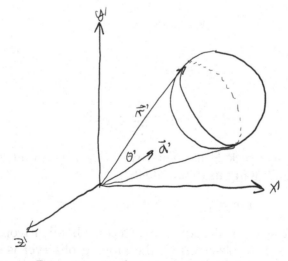

Figure 6.4:   A sphere seen in its rest frame.

at $t_o = 0$ is given by

$$r' \cdot a' = x'a'^x + y'a'^y + z'a'^z = r' \cos \theta' = -ct' \cos \theta', \qquad (6.6)$$

where $t' < 0$ is the time of emission of the light ray. Geometrically speaking the above equation describes a cone. An analogous equation for the light rays reaching another observer moving with velocity V along x, at the moment when both observers overlap, can be obtained by applying the Lorentz transformation (1.7) to (6.6):

$$\frac{x - Vt}{\sqrt{1 - V^2/c^2}} a'^x + ya'^y + za'^z = -c \frac{t - xV/c^2}{\sqrt{1 - v^2/c^2}} \cos \theta'. \qquad (6.7)$$

By multiplying both sides by an arbitrary constant N and rearranging some terms we obtain the final equation:

$$x \frac{a'^x - V \cos \theta'/c}{\sqrt{1 - V^2/c^2}} N + yNa'^y + zNa'^z = -ctN \frac{\cos \theta' - Va'^x/c}{\sqrt{1 - V^2/c^2}}. \qquad (6.8)$$

Let's take a closer look at our result — it has exactly the same form as (6.6), but written with unprimed coordinates and with the redefined a and θ:

$$a^x \equiv N \frac{a'^x - V \cos \theta'/c}{\sqrt{1 - V^2/c^2}},$$

$$a^y \equiv Na'^y,$$

$$\qquad (6.9)$$

$$a^z \equiv Na'^z,$$

$$\cos \theta \equiv N \frac{\cos \theta' - Va'^x/c}{\sqrt{1 - V^2/c^2}}.$$

We can always pick N in such a way that a is a unit vector and $|\cos \theta| \leq 1$, allowing us to rewrite (6.8) as

$$r \cdot a = xa^x + ya^y + za^z = -ct \cos \theta, \qquad (6.10)$$

which, once again, is the equation of a cone! In other words the outline of the sphere observed by the moving observer is still circular — therefore, the moving sphere still appears to be a sphere in spite of Lorentz contraction! Since the position of the sphere in its rest frame is completely arbitrary, our result is general. A relativistically moving sphere will always cast a circular shade. The direction and the size of the shade may change from frame to frame, but it will remain circular. Isn't it a spectacular optical illusion?

6.5 Questions

- Does the speed of light in a river depend on the speed of the water?
- Is light emitted from a source approaching an observer always blue shifted?
- We know that a moving sphere appears to have a spherical shape. Can we use that fact to infer anything about the apparent shape of a moving circle when the circle passes by a camera at the minimum distance? Justify the answer with equation (6.5).

6.6 Exercises

- Consider a light ray of a wavelength λ falling on a mirror at an angle α. Suppose that the mirror is moving perpendicular to its surface with velocity v. Calculate the angle and wavelength of the reflected ray.
- Repeat calculations from the previous exercise for a mirror moving in a direction parallel to its surface.
- A driver is pulled over by a policeman for running a red length. The driver tries to explain that, in his reference frame, the red signal appeared to be green due to the Doppler effect. Can the policeman charge the driver for speeding if the maximum allowed speed in this area is $0.5c$?
- A source of light approaches an inertial observer at a constant speed along such a trajectory that the registered spectrum is the same as for the source when it is at rest. Determine the trajectory that allows this.
- A small object approaches an inertial observer with an apparent radial velocity equal to $2c$. Calculate the real velocity of the object.

- What is the maximum possible apparent speed of a moving object?
- In the example of the moving sphere, prove that if the value of N in the formula (6.9) is chosen such that a is a unit vector then $|\cos\theta| \leq 1$.

Chapter 7

Relativistic Dynamics

7.1 Four-Vectors

So far, we have managed to entertain ourselves with six chapters full of relativistic *kinematics*. In comparison, the topic of non-relativistic kinematics makes for less dazzling cocktail party banter. As non-relativistic kinematics is practically trivial, there is little to discuss on that subject. Soon, we will turn our attention to *dynamics* and eventually derive the famous $E = mc^2$ formula. But before we attach the cart to the horse, we ought to familiarise ourselves with the surprisingly useful notion of a *four-vector*. Understanding this concept will soon allow us to solve the hardest problem in relativistic kinematics (which you may recall from our hair-raising encounter in Section 3.2) in just a single line.

Four-vectors dwelling in a 4D spacetime play a similar role to vectors (also called *three-vectors*) in a 3D space. The most natural four-vector can be composed of four quantities: time t, and three spacial components x, y, and z, characterising some event happening in spacetime. Such a four-vector is called a *four-position*, and it is written as (ct, x, y, z) or (ct, r) for short. The extra c constant has been added so that all of the components share the same physical units. From previous chapters we already know how a four-position transforms between inertial reference frames from (ct, r) into (ct', r'). Lorentz transformation (3.5) characterising this transition keeps the following quantity invariant: $(ct)^2 - r \cdot r$, and for that reason we will call this quantity the "square length" of the four-position. We used quotation marks because this expression can

87

be positive, null, or negative. In accordance with the terminology of spacetime intervals, a four-vector of a positive square length will be called *timelike*, a four-vector of a null length will be called *null*, and when the square length is negative the four-vector will be called *spacelike*.

The simple example of four-position can inspire us to generalise the notion of the four-vector to any four-element object (A^0, A^1, A^2, A^3) or $(A^0, \mathbf{A})$ that, like the four-position, transforms between inertial reference frames according to that same transformation rule. In other words, when we make a transition to a frame moving with velocity V along x, $(A^0, \mathbf{A})$ transforms into $(A'^0, \mathbf{A}')$ given by

$$A'^0 = \frac{A^0 - A^1 V/c}{\sqrt{1 - V^2/c^2}},$$

$$A'^1 = \frac{A^1 - A^0 V/c}{\sqrt{1 - V^2/c^2}},$$

$$A'^2 = A^2, \hspace{4cm} (7.1)$$

$$A'^3 = A^3,$$

or in a more general case of an arbitrary relative velocity V of the moving frame:

$$A'^0 = \frac{A^0 - \frac{\mathbf{A} \cdot \mathbf{V}}{c}}{\sqrt{1 - \frac{V^2}{c^2}}},$$

$$\mathbf{A}' = \mathbf{A} - \frac{\mathbf{A} \cdot \mathbf{V}}{V^2}\mathbf{V} + \frac{\frac{\mathbf{A} \cdot \mathbf{V}}{V^2}\mathbf{V} - A^0\frac{\mathbf{V}}{c}}{\sqrt{1 - \frac{V^2}{c^2}}}. \hspace{1cm} (7.2)$$

We should stress that these are the defining properties of four-vectors. Any random collection of four quantities is not a four-vector, because it does not transform properly — just like a collection of three quantities is not automatically a three-vector. Your date of birth consists of three numbers, but it doesn't change no matter how many times you rotate yourself. Vectors, on the other hand, must change their components under rotation. So, your date of birth is definitely not a vector.

Four-vectors are typically denoted by the Greek superscripts A^μ. By applying the above transformation law (7.2), we can verify that the "square length" of the four-vector A^μ defined as $(A^0)^2 - A \cdot A$ does not change and is equal to $(A'^0)^2 - A' \cdot A'$. This leads to an appealing analogy between the fact that 3D rotations do not affect the lengths of three-vectors, and 4D hyperbolic rotations do not change the lengths of four-vectors. All the physical quantities we are used to, such as energy, momentum, electric charge, time, space, the modulus square of the wave function, a matrix determinant, or even a trace in algebra (just to name a few) are only important because they do not change very easily. All these quantities are invariant under some operations, which is why we like them so much. The length of a four-vector can be written using a very elegant notation involving Greek indices μ and ν that take discrete values from 0 to 3: $\sum_{\mu=0}^3 \sum_{\nu=0}^3 \eta_{\mu\nu} A^\mu A^\nu$, where A^μ is the μth component of the four-vector A^μ, and $\eta_{\mu\nu}$, known as the Minkowski spacetime metric, or *Minkowski metric* for short, is given by

$$\eta_{\mu\nu} = \begin{pmatrix} 1 & 0 & 0 & 0 \\ 0 & -1 & 0 & 0 \\ 0 & 0 & -1 & 0 \\ 0 & 0 & 0 & -1 \end{pmatrix}. \tag{7.3}$$

Usually, we omit the summation symbol when it is involved in mononomial expressions. By default, we perform the summation over the repeating Greek index across the whole range of its possible values. This practice of omitting the summation symbol is called the *Einstein summation convention*. Accordingly, the length of a four-vector is usually written as $\eta_{\mu\nu} A^\mu A^\nu$.

Four-vectors have several useful properties. The most common ones follow from linearity of the Lorentz transformation, which implies that any linear combination of four-vectors, such as $\alpha A^\mu + \beta B^\mu$, is also a four-vector. For that reason the Lorentz transformation does not change the expression: $\eta_{\mu\nu}(\alpha A^\mu + \beta B^\mu)(\alpha A^\nu + \beta B^\nu) = \eta_{\mu\nu}(\alpha A'^\mu + \beta B'^\mu)(\alpha A'^\nu + \beta B'^\nu)$. By substituting the equalities $\eta_{\mu\nu} A^\mu A^\nu = \eta_{\mu\nu} A'^\mu A'^\nu$ and $\eta_{\mu\nu} B^\mu B^\nu = \eta_{\mu\nu} B'^\mu B'^\nu$ we find that not only is the length of four-vectors invariant under the Lorentz transformation, but so is the *scalar product* of two four-vectors given by: $\eta_{\mu\nu} A^\mu B^\nu = \eta_{\mu\nu} A'^\mu B'^\nu$.

7.2   Four-Velocity

How about some examples to make things a little more concrete before we jump into even lesser known and more surprising properties of four-vectors? Consider the rest frame of an airborne Witch. Suppose that a four-vector in this rest frame has the following elements: $(c, 0, 0, 0)$. What if we change the frame to one in which the Witch's velocity equals v? The new frame must move with velocity $-v$, and the corresponding Lorentz transformation (7.2) transforms our four-vector into:

$$v^\mu = \left(\frac{c}{\sqrt{1 - v^2/c^2}}, \frac{v}{\sqrt{1 - v^2/c^2}} \right). \tag{7.4}$$

This particular four-vector parameterised by the Witch's velocity v is called a *four-velocity*. If we decide to jump to another frame moving with velocity V, the Witch's velocity will change from v to v', given by (3.6), and v^μ will Lorentz transform into v'^μ. There are two ways of calculating the four-velocity v'^μ. We can either take the formula (7.4) and replace v with v' given by (3.6) or we can invoke the definition of a four-vector and apply transformation formulas (7.2) to (7.4). Both methods yield identical results:

$$v'^\mu(v) = v^\mu(v'). \tag{7.5}$$

To verify that, let's backtrack to the expression (3.15) and rewrite it in the following form:

$$\frac{c}{\sqrt{1 - v'^2/c^2}} = \frac{c - v \cdot V/c}{\sqrt{1 - v^2/c^2}\sqrt{1 - V^2/c^2}}. \tag{7.6}$$

Note that the left hand side of (7.6) is the "temporal" component of v'^μ obtained by replacing v with v' in (7.4). The right-hand side is the same "temporal" component of v'^μ, but obtained by the direct application of the Lorentz transformation (7.2) to the four-vector (7.4). The equality (7.6) is a proof that both methods yield identical outcomes. The same can be shown for the remaining "spacial" part of the four-vector (7.4). By multiplying the formulas (3.6) and

(7.6) by sides, we obtain

$$\frac{v'}{\sqrt{1 - v'^2/c^2}} = \frac{\sqrt{1 - V^2/c^2}\left(v - \frac{v \cdot V}{V^2}V\right) - \left(V - \frac{v \cdot V}{V^2}V\right)}{\sqrt{1 - v^2/c^2}\sqrt{1 - V^2/c^2}}, \qquad (7.7)$$

which shows the equality of results obtained by both approaches for the remaining component of the four-velocity. This completes the proof of (7.5).

Another thing worth grasping about the four-velocity is the fact that its length does not depend on the choice of the observer:

$$\left(\frac{c}{\sqrt{1 - v^2/c^2}}\right)^2 - \left(\frac{v}{\sqrt{1 - v^2/c^2}}\right)^2 = c^2. \qquad (7.8)$$

Indeed, that length is simply c^2 regardless of the velocity v. Finally, let us note that the four-velocity v^μ is nothing more than a derivative of four-position $x^\mu = (ct, r)$ over proper time $d\tau = \sqrt{1 - v^2/c^2}dt$:

$$v^\mu = \frac{dx^\mu}{d\tau}. \qquad (7.9)$$

The infinitesimal proper time, $d\tau$, is a relativistic invariant because it is proportional to the spacetime interval:

$$d\tau = \sqrt{1 - v^2/c^2}\, dt = \frac{1}{c}\sqrt{c^2\, dt^2 - dr^2} = \frac{1}{c}ds. \qquad (7.10)$$

As a consequence, the derivative (7.9) is still a four-vector.

7.3 One-Line Solution to the Hardest Problem in Special Relativity

Withour further ado, let's move on to a secret property of four-vectors that can be extremely useful and will give us an advantage in dealing with some hard 3D problems.

Consider an inertial unprimed reference frame, denoted with K, and a primed one, called K', moving with the relative velocity V. We will show that, for an *arbitrary* four-vector A^μ, the following identity always holds:

$$\frac{A - A'}{A^0 + A'^0} = \frac{V/c}{1 + \sqrt{1 - V^2/c^2}}. \tag{7.11}$$

The left-hand side involves a ratio between the difference of spacial components and the sum of temporal components taken in the two frames K and K'. The right-hand side does not depend on the four-vector A^μ, only on the relative velocity between frames. That seems peculiar. In particular, for *arbitrary* four-vectors A^μ and B^μ the following equality must hold:

$$\frac{A - A'}{A^0 + A'^0} = \frac{B - B'}{B^0 + B'^0}. \tag{7.12}$$

In order to prove the rule (7.11) we'll introduce the third, double-primed reference frame K'', in which K' moves with some velocity U, while K is moving the velocity $-U$ as shown in Fig. 7.1. Therefore, the newly introduced K'' frame is "in between" K and K' in terms of the relative velocity. Since all the relative motions are collinear, we can assume that all the frames' axes are non-rotated. Let us use the transformation formulas (7.2) to transform both of the four-vectors A^μ and A'^μ appearing in the left-hand side of equation (7.11), to the frame K''. In order to transform A^μ to the frame

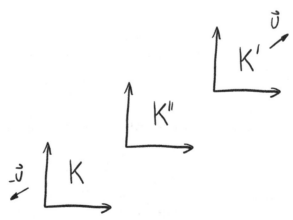

Figure 7.1:   Reference frames K and K' in a relative, collinear motion, as seen from the auxiliary inertial frame K''.

K'' we need to substitute the velocity $V \rightarrow -U$, $A'^{\mu} \rightarrow A^{\mu}$, and $A^{\mu} \rightarrow A''^{\mu}$ in the formulas (7.2) and to transform A'^{μ} we have to substitute $V \rightarrow U$ and $A^{\mu} \rightarrow A''^{\mu}$ in (7.2). As a result we obtain that $A - A' = \frac{2A^{0''}U/c}{\sqrt{1-U^2/c^2}}$, as well as $A^0 + A^{0'}\frac{2A^{0''}}{\sqrt{1-U^2/c^2}}$ and the whole left-hand side of (7.11) reduces to a very simple expression: $\frac{u}{c}$. Using the velocity transformation formula (3.6) for $v \rightarrow U$, $V \rightarrow U$ and $v' \rightarrow V$ we also obtain that U and V are related via the relation $V = \frac{2U}{1+U^2/c^2}$, which can be easily inverted yielding the right-hand side of equation (7.11).

Now that we have proven the property (7.11), solving the hardest problem in relativistic kinematics from Section 3.2 will be totally effortless. Remember how much work we needed to derive the formula (3.17)? Now we can see that this formula is just a special case of our newly discovered property (7.11) for A^{μ} replaced with the four-velocity v^{μ} given by (7.4). Just think of how much work we could have saved, if only we had known the notion of a four-vector in advance. But more good things are about to come thanks to our newly discovered tools.

7.4 Energy and Momentum

Now it's time for the serious stuff. We will look for relativistic expressions characterising energy and momentum of moving objects. Imagine a scenario, in which a closed system contains a number of objects with initial momenta p_i and energies E_i. After some time has passed and these object have been colliding and interacting through some energy-conserving processes, they end up with final momenta $\tilde{p}_j$ and energies $\tilde{E}_j$. Conservation laws require that:

$$\sum_i p_i - \sum_j \tilde{p}_j = 0,$$
$$\sum_i E_i - \sum_j \tilde{E}_j = 0. \tag{7.13}$$

But wait, there's more! The above condition has to be satisfied in *all* possible inertial frames, not only in this one. When we change the frame of reference, the momenta and energies should transform accordingly so that, if the condition (7.13) is satisfied in one frame, it is automatically satisfied in all other frames.

Let us make a crucial point. Suppose that the energy and momentum of an object formed a four-vector structure $p^\mu \equiv (E/c, p)$. In this case, the conservation laws (7.13) could be written as

$$\sum_i p_i^\mu - \sum_j \widetilde{p}_j^\mu = (0, 0), \qquad (7.14)$$

which means that the above linear combination of four-vectors must be a null four-vector, and all its elements vanish. And, seeing as a Lorentz transformed version of such a four-vector also vanishes, both energy and momentum would automatically be conserved in all other frames. Bingo!

It transpires that all we need to do is to look for any velocity-dependent four-vector, that is conserved in at least one frame of reference. It will already guarantee the validity of conservation laws in all other frames. Shall we try something simple? Let's define the *four-momentum* p^μ by taking the four-velocity v^μ and multiplying it by the mass m of a given body:

$$p^\mu \equiv mv^\mu = \left(\frac{mc}{\sqrt{1 - v^2/c^2}}, \frac{mv}{\sqrt{1 - v^2/c^2}} \right). \qquad (7.15)$$

This new object has the correct non-relativistic limit, because for small velocities it reduces to known expressions for the energy and momentum of a free body:

$$E \equiv \frac{mc^2}{\sqrt{1 - v^2/c^2}} \approx mc^2 + \frac{mv^2}{2} + \cdots,$$

$$p \equiv \frac{mv}{\sqrt{1 - v^2/c^2}} \approx mv + \cdots \qquad (7.16)$$

so, we may be on the right track. An extra constant factor mc^2 in the expansion of the energy of the free body appears to have snuck in, but non-relativistic energy is defined up to an additive constant anyway, so it needn't vex us. The only question that remains is the

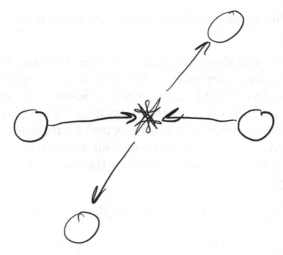

Figure 7.2: Two identical balls collide elastically in the centre-of-mass frame.

following: can this four-vector be conserved in at least one reference frame? Let's find out.

We first consider an *elastic* collision of two identical balls of mass m and initial velocities $\pm v$, observed in the centre-of-mass reference frame — Fig. 7.2. What will be the balls' velocities after collision? We are not allowed to infer the answer from the con-servation laws, because our goal is to prove these laws. We can, however, invoke the symmetry argument. What do we mean by "elastic" collision anyway? Typically, we'd identify elastic collisions by the fact that they conserve the mechanical energy, but for the pur-pose of this discussion, a pertinent characteristic is that they are also *time-reversible*.

It follows from this definition that the magnitude of their velocities will not change after the collision. If the final velocities were decreased, then for the time-reversed process the velocities would have to increase. As a consequence, some initial velocities would increase in the process while others would decrease. The existence of some preferred velocity of the balls, that does not change during the collision, would also be inevitable. Since no such preferred velocity exists, we must conclude that elastic collisions in the centre-of-mass frame always preserve the initial velocities. This in turn means that, in elastic collisions of identi-cal balls, the total energy, as well as the total momentum defined through (7.16), is preserved. These two are equal to $\frac{2mc^2}{\sqrt{1-v^2/c^2}}$ and 0, respectively.

How about collisions between bodies of different sizes? In this case, the symmetry argument cannot be used. However, we can always argue that all objects are made of identical atoms and any macroscopic collision is in fact a chain of two-atom elastic interactions that always preserve both energy and momentum.

The length of the four-momentum four vector is constant, which leads to the following relation between the energy and momentum of a body of mass m:

$$E^2 = (mc^2)^2 + p^2c^2. \tag{7.17}$$

Interestingly, for massless particles the above formula reduces to the relation $E = pc$. Notice that if objects of zero mass, like photons, are to carry non-zero energy, then according to the definition (7.16) their speed must be always equal to c.

We can also answer another important question: why is it not possible to accelerate any massive object beyond the speed of light? The answer lies in the definition of energy (7.16). In order to reach the speed of light we would have to use infinite amount of energy, which is physically impossible. However, this does not indicate that superluminal particles cannot exist. All we are stating is that subluminal particles cannot be accelerated beyond the speed of light limit.

7.5 $E = mc^2$

This may come as a shock, but we're finally ready to discover the origin of the Einstein's legendary formula $E = mc^2$. It'll all click together by the end of this section. Let's start off by reflecting on the zero-order term mc^2 present in the expansion of energy (7.16). Although non-relativistic energy is defined up to an additive constant, this particular constant is necessary for energy and momentum to form the four-vector structure. And that structure turned out to be crucial for the relativistic invariance of the conservation laws. So, is there any deeper physical meaning to this constant term? Let us find out.

In Section 7.4, we verified that energy and momentum are preserved in all elastic collisions in all inertial frames. But what about collisions that are *inelastic*? Let's imagine a perfectly inelastic collision of two identical balls of mass m, moving in the opposite directions with velocities $\pm v$ in their centre-of-mass frame. Suppose the collision makes the balls stick together and form a single resting body of mass $2m$. The total momentum of the system, both before and after the collision, vanishes. So, the momentum is conserved. But the energy is not, and in a non-relativistic approach this is usually attributed to an extra heat Q produced in the collision.

Unfortunately, the above scenario is completely unacceptable in a relativistic theory! And the reason will soon be clear. An inelastic collision considered in the centre-of-mass frame can be characterised by the updated formula (7.14) that now takes the form:

$$\sum_i p_i^\mu - \sum_j \widetilde{p}_j^\mu = (Q/c, 0), \tag{7.18}$$

where $Q > 0$. The problem becomes obvious once we change the reference frame. Lorentz transformation of the four-vector (7.18) will transform the right-hand side, $(Q/c, 0)$, into a four-vector that has a non-vanishing spacial component. This means that, in the new frame, the momentum will not be conserved! This is unacceptable! We must have made a mistake, because momentum must be conserved in all reference frames for all types of collisions.

The reason for this breakdown is the fact that mechanical energy appears not to be conserved in inelastic collisions and, as a consequence, momentum is not conserved either. It seems there's no way to fix it except to find the way that energy can be conserved even in inelastic processes. Does that mean we need to find a relativistic description of heat and take it into account in order to rescue the conservation of energy? Not at all — there is another, brilliant solution.

The total energy before the collision is equal to $2mc^2/\sqrt{1 - v^2/c^2}$, which is larger than $2mc^2$. If energy is to be conserved then the energy after the collision must also be $2mc^2/\sqrt{1 - v^2/c^2}$. Since the product of the collision is resting, its mass has to *increase*! In order for the total energy to be conserved, the new mass should be equal to $2m/\sqrt{1 - v^2/c^2}$ instead of $2m$. But is it possible for an inelastic collision to change the total mass of the system? After all, the number of atoms before and after the collision is exactly the same!

Don't forget that material heats up when it collides, which means that all the atoms of the material start to move faster. So, is a moving object heavier than one at rest? Well, if we heat up a potato and place it on a very precise scale, indeed we will see that it has become slightly *heavier*. Taking this into account, one could say that the two colliding balls do not weight $2m$, but rather $2m/\sqrt{1 - v^2/c^2}$, which, consequently, is also the weight of the resulting, merged ball.

In particle physics, the situation is usually defined in a slightly different manner. This is because, when two elementary particles without any internal structure collide producing another elementary particle, the mass of the product is increased in such a way that the total energy is conserved. Therefore, particle physicists introduce the notion of a *rest mass* of an elementary particle as the only "real" mass, while everything non-elementary has a "mass" that consists of the rest masses of all the ingredients, plus all the extra internal energies (divided by c^2). As a result, whenever a particle physicist buys potatoes at the grocery store, he should measure their weight in Joules rather than kilograms. This is because, according to this particle paradigm, elementary particles are the only objects that have well-defined masses. Unfortunately, the theory of relativity says nothing about whether matter consists of elementary particles or not, making it awkward to justify this approach using bare theory of relativity.

On the other hand, if we decide to attribute mass to any macroscopic objects, then we have no way of avoiding the concept of velocity-dependent "relativistic" mass. A balloon filled with a noble gas containing N freely moving molecules will change its total mass (inertial and gravitational) if we heat the gas up, i.e. make the gas molecules move faster. Since the mass of the whole is N times the (average) mass of a single moving gas molecule (plus the mass of the shell) then clearly the mass of a single molecule must increase with its velocity.

To some degree this is only semantics, but what is certain is the following. If we have two objects of a rest mass m and want to produce another object of a larger mass, we need to pump some energy in. For example if we accelerate both objects to velocity v, we have to spend the energy cost equal to

$$\Delta E = \frac{2mc^2}{\sqrt{1 - v^2/c^2}} - 2mc^2. \tag{7.19}$$

This will allow us to produce extra mass in a non-elastic collision of the accelerated objects, equal to

$$\Delta m = \frac{2m}{\sqrt{1 - v^2/c^2}} - 2m, \tag{7.20}$$

which results in the relation:

$$\Delta E = \Delta mc^2, \tag{7.21}$$

stating that the energy can be converted into mass, and vice versa. This is a prerequisite for the universal conservation of momentum. Notice that the mysterious additive zero-order term mc^2 appearing in (7.16) acquired a novel interpretation. It corresponds to the so-called *rest energy* stored within any massive object at rest, that could potentially be turned into any other type of energy. Ever heard of atomic bombs? The principle of their action is simple: masses of their uranium components are decreased creating immense amounts of energy. Relativistic energy is not defined only up to an additive constant anymore. There exists an absolute scale and we can drop the "Δ" symbols in the above equation, which brings us to the famous formula:

$$E = mc^2, \tag{7.22}$$

matching the title of this section.

7.6 May the (Relativistic) Force Be With You

For any closed system of bodies the total energy and the total momentum are conserved. Of course, the momentum of an individual body can change. Whenever that happens, we say that some *force* acted on the body. The notion of force is in fact only ancillary — it can be a convenient way of describing interactions between bodies. In non-relativistic mechanics, force appears in Newton's Second Law (which in fact is not a law, but rather a definition of force) as $F = ma$ or, equivalently, $F = \frac{dp}{dt}$. How may we generalise the notion of force to the relativistic case? Let us note that in special relativity $ma = m\frac{dv}{dt}$ is not the same as $\frac{dp}{dt} = m\frac{d}{dt}\frac{v}{\sqrt{1-v^2/c^2}}$. Not only do these vectors have different lengths, but they can also have different directions. Therefore, in order to define a relativistic force, we must decide, which of these alternative definitions we will choose. Unlike acceleration, momentum plays an important physical role in the evolution of dynamical systems (after all its total amount is conserved in closed systems). Therefore, relativistic force is defined as

$$F = \frac{dp}{dt}. \tag{7.23}$$

Of course, we can always extend the considered system, so that it contains all interacting bodies, in which case the change in the total momentum has to vanish. In particular, for a pair of interacting bodies this results in Newton's third law. We'll return to the subtleties of that law later on, when we discuss electromagnetic interactions in more detail.

Since the time interval dt depends on the observer, one can expect that the force will also be frame dependent. What kind of dependence is it? We already stumbled upon this problem in Section 2.5, when we discussed the force of attraction between an electric charge and a straight wire conducting electric current. We realised that all paradoxes disappear if we assume that the force F', acting in a frame moving together with the electronic current V, is related to the force F in the rest frame of the cable via the formula:

$$F' = \frac{F}{\sqrt{1 - V^2/c^2}}. \tag{7.24}$$

This suggests that relativistic force is a component of some four-vector. We will leave the task of a more detailed investigation to an interested reader.

7.7 Massless Particles and Planck's Postulate

So far, we have described massive particles. But what about light? After all, light carries both energy and momentum, but has no mass. A quick look at equation (7.17) reveals that, for massless objects, the following equation must hold: $E = pc$. A detailed study of Maxwell's equations and the Poynting theory of electromagnetic fields fully supports this conclusion. We will return to this topic later. For now, let us consider a frame of reference, in which a portion of light of energy E and momentum $p = \frac{E}{c}$ is emitted along the x axis by a resting source. An energy–momentum four-vector of such a portion has the form $p^\mu = \frac{E}{c}(1, 1, 0, 0)$. Let us transform this four-vector to a frame, in which the source is moving with velocity v along x:

$$p'^\mu = \frac{E}{c}\left(\sqrt{\frac{1 + v/c}{1 - v/c}}, -\sqrt{\frac{1 + v/c}{1 - v/c}}, 0, 0\right). \tag{7.25}$$

This means that the energy of a portion of light depends on the velocity of the source proportionally to the coefficient $\sqrt{\frac{1+v/c}{1-v/c}}$. By recalling our conclusions from Section 6.1, we should notice that it is exactly the same dependence as in the 1D case of the Doppler effect characterising the effect of motion on the frequency of light.

Such a simple observation can lead to a groundbreaking conclusion. When Max Planck postulated the existence of *quanta* of light, he assumed that the energy of such quanta is proportional to the frequency v of the corresponding electromagnetic wave:

$$E = hv. \tag{7.26}$$

Would it be legal to postulate any other type of dependence? Our recent conclusion shows that any function of frequency, other than linear, would transform differently than energy. It means that only the linear relation between energy and frequency of light is relativistically invariant. Even the slightest modification of Planck's hypothesis would result in conflict with relativity. In 1900, Planck could not have known about relativity yet. But it should come as no surprise that, five years later, Einstein was the first to take Planck's postulate seriously.

7.8   Superluminal Particles

By now we have characterised energy–momentum four-vectors of both massive and massless objects (light). Our methodology was rather straightforward. For massive particles we first introduced a time-like four-vector $mc(1, 0)$ characterising the energy and momentum of a resting particle and then applied the Lorentz transformation (7.2) to a frame, in which the particle was moving with velocity v. This resulted in

$$p^\mu = mc\left(\frac{1}{\sqrt{1 - v^2/c^2}}, \frac{v/c}{\sqrt{1 - v^2/c^2}}\right). \qquad (7.27)$$

The same method was applied to massless objects, such as portions of light. Since these objects always move with the same speed and have no rest frame, it was natural to first describe them in a frame in which their source (or any other reference object) is resting. In this frame, their energy and momentum null four-vector has the form $\frac{E}{c}(1, s)$, where E is the energy of light and the unit vector s characterises the direction of propagation. In order to determine the four-momentum in a frame in which the source is moving with an arbitrary velocity v, we apply the Lorentz transformation (7.2) yielding:

$$p^\mu = \frac{E}{c}\left(\frac{1 + \frac{s \cdot v}{c}}{\sqrt{1 - v^2/c^2}}, s - \frac{s \cdot v}{v^2}v + \frac{\frac{s \cdot v}{v^2}v + \frac{v}{c}}{\sqrt{1 - v^2/c^2}}\right). \qquad (7.28)$$

When the motion takes place along the direction of propagation of light, our result simplifies to the more familiar:

$$p^\mu = \frac{E}{c}\left(\sqrt{\frac{1 + v/c}{1 - v/c}}, \sqrt{\frac{1 + v/c}{1 - v/c}}s\right). \qquad (7.29)$$

Now it is time to move to a non-orthodox case of superluminal bodies, briefly introduced in Section 4.3. We know that the presence of such objects introduces non-classical phenomena involving non-deterministic behaviour and the possibility of travelling along multiple paths at once. How about we ask a more down-to-earth question. What are the expressions for the energy and momentum of such hypothetical, exotic objects? In order to discover these expressions, we will follow the exact same procedure as the one summarised above. Beware, however, because the next few pages will be the most technical part of this book.

We first notice that superluminal objects have no rest-frame within the family of subluminal observers that we consider here. This follows directly from equation (3.15): if an object is superluminal in one inertial frame, then it is also superluminal in any other (subluminal) frame of reference. There exists, however, a frame in which a superluminal object moves infinitely fast towards some direction s. We will assume that in this frame, the energy–momentum characterising such an object of a mass m is a space-like four-vector (otherwise we return to the familiar case of subluminal objects or light). This suggests that the energy of the superluminal object in this frame vanishes, and we will choose the momentum to be along the direction of motion, s: $mc(0, s)$. Notice that this situation mirrors the scenario involving an orthodox subluminal body, in which case its momentum vanished in the rest frame of that body. The two scenarios differ by interchanging the time-like and space-like four-vector components, just like we have witnessed in (4.14).

We will be interested in determining the expression for the energy and momentum of the superluminal body in a frame, in which the object is moving with an arbitrary superluminal velocity v. This is no easy task! For subluminal objects, we only had to Lorentz transform the rest-frame expression to the frame that moves with velocity $-v$. This time, we should make a serious effort to find the appropriate velocity $-V$ of the new frame. Finding it will allow us to apply the Lorentz transformation (7.2) for that relative velocity $-V$ that will give us the energy–momentum four-vector:

$$p^\mu = mc\left(\frac{\frac{s \cdot V}{c}}{\sqrt{1 - V^2/c^2}}, s - \frac{s \cdot V}{V^2}V + \frac{\frac{s \cdot V}{V^2}V}{\sqrt{1 - V^2/c^2}}\right). \quad (7.30)$$

Here, the unknown frame velocity $-V$ is such that it transforms an infinite velocity along s into a superluminal velocity v according to equation (3.6).[a] Let's look more closely at the velocity transformation formula (3.6) for that case. We need to take that formula and substitute $v' \to v$, $V \to -V$, and $v \to \infty s$, where the infinity symbol indicates taking the infinite limit. This procedure yields the following equation:

$$\frac{s \cdot V}{c^2} v = \sqrt{1 - \frac{V^2}{c^2}} \left(s - \frac{s \cdot V}{V^2} V \right) + \frac{s \cdot V}{V^2} V, \qquad (7.31)$$

which upon substituting into (7.30) leads to

$$p^\mu = mc \left(\frac{\frac{s \cdot V}{c}}{\sqrt{1 - V^2/c^2}}, \frac{\frac{s \cdot V}{c^2} v}{\sqrt{1 - V^2/c^2}} \right). \qquad (7.32)$$

In order to determine the unknown velocity V, we will first take the scalar product of (7.31) with the versor s. Since the right-hand side of the resulting equation is always positive, we obtain the following condition:

$$(s \cdot V)(s \cdot v) > 0, \qquad (7.33)$$

which is equivalent to $\mathrm{sgn}(s \cdot V) = \mathrm{sgn}(s \cdot v)$. Next, we'll calculate the scalar product of (7.31) with itself:

$$\frac{(s \cdot V)^2}{c^4} v^2 = \left(1 - \frac{V^2}{c^2} \right) \left(s - \frac{s \cdot V}{V^2} V \right)^2 + \frac{(s \cdot V)^2}{V^2}$$

$$= 1 - \frac{V^2}{c^2} + \frac{(s \cdot V)^2}{c^2}. \qquad (7.34)$$

After moving the last term on the right-hand side to the left-hand side of the equation, we take the square root of both sides and use

---

[a]Not all pairs of s and v can be related this way. It can be shown that the following condition must be satisfied: $v^2 - c^2 < (s \cdot v)^2$.

the condition (7.33), yielding:

$$\frac{s \cdot V}{c} = \frac{\sqrt{1 - V^2/c^2}}{\sqrt{v^2/c^2 - 1}} \, \text{sgn}(s \cdot v). \tag{7.35}$$

We can finally substitute the resulting equation (7.35) into (7.32) obtaining the expression for the four-momentum of a superluminal object:

$$p^\mu = mc \left(\frac{\text{sgn}(s \cdot v)}{\sqrt{v^2/c^2 - 1}}, \frac{\text{sgn}(s \cdot v)\frac{v}{c}}{\sqrt{v^2/c^2 - 1}} \right). \tag{7.36}$$

It becomes clear now, why superluminal objects cannot move with subluminal speeds. Since their energy increases with decreasing velocity, it would take an infinite amount of energy to slow them down under the speed of light. So, while special relativity does not deny the existence of superluminal objects, it certainly denies the possibility of any object, whether massive or massless, crossing the speed of light in any direction.

It is beyond interesting, that the resulting expressions (7.36) for the energy and momentum of the superluminal object depend not only on its velocity v, but also on the direction s. We should also notice that the term depending on that direction, $\text{sgn}(s \cdot v)$, has non-trivial Lorentz transformation properties due to Thomas–Wigner rotation. Let's do some more detective work, gumshoes! Suppose that we want to apply another Lorentz transformation corresponding to some subluminal velocity $\widetilde{V}$ to the four-momentum (7.36). We can either transform the superluminal velocity appearing in (7.36), $v \rightarrow v'$ using the formula (3.6) as well as the versor $s \rightarrow s'$, or simply Lorentz transform the whole four-vector using (7.2). The result should be the same. For the time-like component we have

$$\frac{\text{sgn}(s' \cdot v')}{\sqrt{v'^2/c^2 - 1}} = \frac{\text{sgn}(s \cdot v)}{\sqrt{v^2/c^2 - 1}} \frac{1 - \frac{v \cdot \widetilde{V}}{c^2}}{\sqrt{1 - \widetilde{V}^2/c^2}}. \tag{7.37}$$

Using (3.15) in the above equation we obtain

$$\text{sgn}(s' \cdot v') = \text{sgn}(s \cdot v) \, \text{sgn}\left(1 - \frac{v \cdot \widetilde{V}}{c^2}\right). \tag{7.38}$$

Our result (7.38) has a very simple interpretation. Energy and momentum (7.36) of an object moving with a superluminal velocity v are defined up to a sign sgn $(s \cdot v)$. But whatever sign we pick, it must flip whenever a Lorentz transformation is applied with the velocity V such that $v \cdot V > c^2$.

Notice that the formulas (4.12) allow us to transform the four-vector (7.36) to the superluminal, comoving reference frame. It is easiest to start from a subluminal frame, in which the superluminal object moves infinitely fast, so that its four-vector (7.36) is simply $p^\mu = mc(0, s)$. In this case the corresponding superluminal Lorentz transformation (4.12) reduces to (4.14) and yields:

$$p'^\mu = mc(s, 0). \tag{7.39}$$

Keep in mind that, in the superluminal reference frame, spacetime has three time-like components and only one space-like component. So, a single momentum component vanishes and a triple of energies does not. Superluminal objects are interesting animals, aren't they?

7.9 When the Forbidden Becomes Permissible

Relativistic momentum conservation law demands that in all types of collisions, elastic or not, the total energy and momentum are always conserved. Let us discuss a few fine examples.

It is forbidden for a free electron to absorb a photon. To understand why, consider a hypothetical process in the centre-of-mass frame of reference — see Fig. 7.3(a). In such a frame, after the hypothetical process is over, we have a resting electron. Before that absorption, however, we had a moving electron and a moving photon. Clearly, such a process violates conservation of energy, since a moving electron already carries more energy than the one at rest — not to mention the additional energy of the photon to be accounted for.

It is forbidden for the annihilation of an electron and positron pair to produce just a single photon. Again, let us consider this hypothetical process in the centre-of-mass frame of reference — see Fig. 7.3(b). Such a frame can be introduced for an electron and positron pair, but it cannot exist for a single photon that appears after annihilation, because photons have no rest frames.

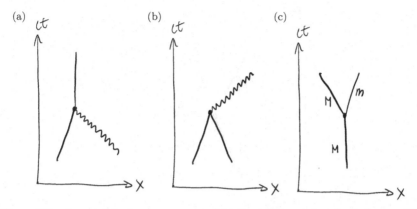

Figure 7.3: Impossible decay processes: (a) a free electron absorbs a photon, (b) an electron and a positron annihilate each other, producing a single photon, and (c) a massive particle M emits another particle m without changing mass.

It is also forbidden for any massive particle to emit another massive particle without a change in mass. The reason is similar to the one in the first of our examples. Just consider the rest frame of the system: before decay we have a resting particle, but after decay we have the same particle in motion plus another emitted particle, as shown in Fig. 7.3(c). That would clearly violate conservation of energy. The rest mass cannot be created only from another particle's kinetic energy.

It turns out, however, that the last process is actually *allowed* — when the emitted particle is superluminal. Consider a frame in which the emitted superluminal particle is moving infinitely fast, and therefore carries only momentum and no energy — as depicted in Fig. 7.4. If the subluminal particle changes its velocity's sign at the moment of emission, then its energy is still the same and the total energy is conserved. The momentum can also be conserved, if the recoil of the subluminal particle balances out with the momentum carried away by the superluminal particle. That can be made if the particle masses are chosen in the right way. Therefore, in principle it is possible for a single, subluminal particle to emit an infinite number of infinitely fast moving particles *without* changing its original mass.

In the above examples, we have considered specially chosen reference frames. Nevertheless, the conservation of energy and

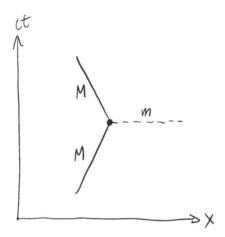

Figure 7.4:   A possible process: a massive particle M emits a superluminal particle m without changing mass.

momentum in one frame implies their conservation in any other inertial frame. For this reason, the above results are completely general.

7.10   Questions

- Can we increase the rest mass of a body by heating it up? What about its weight?
- Must the inelastic collisions conserve the mechanical energy of the colliding objects, or is this only the case for elastic collisions? Is the answer the same in non-relativistic physics, and why?
- When does the product of mass and acceleration of a body have a different direction than the change of its momentum?
- Why is the energy of a single photon proportional to its frequency?

7.11   Exercises

- Determine the rate of precession of Mercury's orbit around the Sun by assuming that its gravitational mass, that appears in the Newton's law of universal gravitation, depends on the velocity according to the formula: $m(v) = \frac{m}{\sqrt{1-v^2/c^2}}$. Ignore all the other relativistic effects.

- Show that a free electron cannot absorb a photon.
- Show that a free electron–positron pair cannot annihilate into a single photon.
- Show that a massive particle cannot emit any other massive particle without changing its rest mass.

Chapter 8

Non-Inertial Frames

8.1   The Clock Postulate — Reexamined

Over seven fat chapters, we have repeatedly discussed inertial frames. A non-inertial frame was only mentioned once, when we stumbled upon the twin paradox. It is therefore time to move on to the next course; a marvellous cuisine of accelerated motions and non-inertial observers. In this Chapter, we will explore amazing phenomena, such as *event horizons* where time comes to a complete stop.

At the end of Section 7.2, we noticed that the four-velocity (7.4) is simply a derivative of the four-position over the proper time $d\tau$. Since proper time is a relativistic invariant, we can obtain "higher-order" four-vectors by calculating higher-order derivatives of the four-position over the proper time. For instance, the second derivative is called a four-acceleration and has the following form:

$$a^{\mu} \equiv \frac{d^2 r^{\mu}}{d\tau^2} = \left(\frac{a \cdot v/c}{(1 - v^2/c^2)^2}, \frac{a(1 - v^2/c^2) + (a \cdot v)v/c^2}{(1 - v^2/c^2)^2} \right), \quad (8.1)$$

where $a \equiv \frac{dv}{dt}$. This new four-vector can be very useful in the study of uniformly accelerated motion.

As we have already pointed out, proper time along a given path connecting two points in spacetime, A and B:

$$\tau = \int_A^B dt \sqrt{1 - v^2(t)/c^2} \quad (8.2)$$

111

is the same in all frames of reference, because it is an integral over infinitesimal spacetime intervals. According to the clock postulate discussed in Section 2.6 an ideal clock is a hypothetical device that measures the proper time dependent on an instantaneous velocity only and is insensitive to any accelerations. We have also invoked results showing that, according to the rules of quantum field theory [5,6] no ideal clocks can exist. This means that any realistic time measuring device will deviate from the perfect formula (8.2) and any corrections must depend on higher derivatives of velocity. Naturally, time measurements with realistic clocks must be the same in all reference frames, so these higher order corrections must be built from relativistic invariants. Consider an invariant that depends not only on instantaneous velocity, but also acceleration. In a situation, when the velocity v is parallel to the acceleration $a = \frac{dv}{dt}$, the four-acceleration (8.1) reduces to

$$a^\mu = \frac{d^2 r^\mu}{d\tau^2} = \left(\frac{av/c}{(1 - v^2/c^2)^2}, \frac{a}{(1 - v^2/c^2)^2} \right), \qquad (8.3)$$

and its square length, multiplied by the squared proper time, is given by

$$\eta_{\mu\nu} a^\mu a^\nu d\tau^2 = -\frac{a(t)^2}{(1 - v^2/c^2)^2} dt^2. \qquad (8.4)$$

By construction, it is still a relativistic invariant. Unsurprisingly, an integral of a square root of that invariant is also a relativistic invariant:

$$\int \frac{a(t)/c}{1 - v(t)^2/c^2} dt = \text{const.} \qquad (8.5)$$

Moreover (and this one is interesting), it has no physical units, it is *dimensionless*. Along with infinitely more invariants constructed in a similar fashion and depending on higher derivatives of velocity, it can be used to modify the expression (8.2) to account for effects of acceleration on real clocks.

Our goal in this chapter is to establish the laws of physics within uniformly accelerated frames. We need accelerating clocks for this, but we have just argued that no realistic clock is ideal. Having said that, we'll just ignore this fact and fudge a little, making an approximation that all our clocks are in fact ideal and measure the proper time along their accelerated paths. Fair enough; in fact, all clocks considered in *general relativity* rely on this approximation.

8.2 Uniformly Accelerated Motion

Is there such a thing as an infinite uniformly accelerated relativistic motion? We are certainly aware that the standard definition involving a constant rate of acceleration is impossible. This is because no body can accelerate without limits, especially beyond the speed of light. Such uniformly accelerated motion only makes sense so long as the velocity is small in comparison to the speed of light.

In order to introduce uniform and relativistic accelerated motion, we'll need to first activate the organ located behind the nose. Let us *define* a uniformly accelerated relativistic motion as one in which a body accelerates at a constant rate in every *instantaneously* co-moving inertial frame. At any instant, one can find a frame in which the accelerating body is instantaneously at rest. A moment later, when the velocity increases, we can find a new frame, and then another, and in any of these frames the rate of acceleration, called *proper acceleration* must be the same[a] — see Fig. 8.1.

Now, we'll ascertain how a uniformly accelerated rocket appears to a fixed, inertial observer. As we have highlighted, at any point of such a motion we can introduce an inertial frame instantaneously co-moving with the rocket, and the rocket will accelerate the same way in every frame. From such a definition, it follows that the whole uniformly accelerated trajectory must be the same to all inertial observers. In other words, the uniformly accelerated trajectory must remain *unchanged* under any Lorentz transformation.

This is rather unintuitive, because in general, any trajectory changes its shape when we observe it from a different frame. Of course, there are the light cones that do not change, because light always moves with the same speed in all inertial frames.

---

[a]We will use the world "proper" to characterise quantities, such as *proper time, proper distance, proper acceleration* etc., that are being measured in an instantaneously co-moving inertial frame of reference.

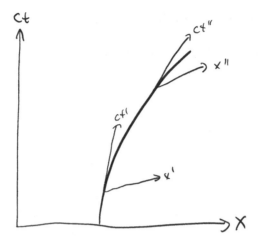

Figure 8.1:   In a uniformly accelerated motion every instantaneously co-moving inertial observer witnesses the motion with the same proper acceleration. These observers have their temporal axes ct' and ct'' tangent to the trajectory.

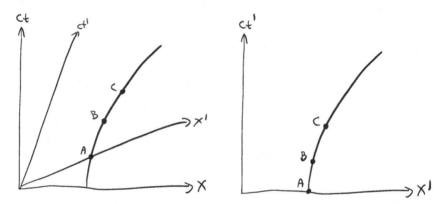

Figure 8.2:   A uniformly accelerated trajectory does not change its shape under any Lorentz transformation, but specific points of that trajectory move across it from frame to frame.

But, according to our definition, a uniformly accelerated trajectory must share this property with the light cones — it has to remain the same under any Lorentz transformation. Individual points of the trajectory can move around, but the whole trajectory has to stay intact — see Fig. 8.2. And this property must hold, no matter what velocity we choose.

We have absorbed enough information to determine the exact shape of the trajectory of the rocket, because we know that one quantity that remains invariant under the Lorentz transformation is $c^2t^2 - x^2$ (if we consider a 1D motion along x). Therefore, the uniformly accelerated trajectory should involve some function f of that invariant, and satisfy the following equation:

$$f\left(c^2t^2 - x^2\right) = \text{const.} \tag{8.6}$$

Taking the inverse f of both sides we obtain

$$c^2t^2 - x^2 = f^{-1}\,(\text{const}) \equiv -K, \tag{8.7}$$

where K is some arbitrary constant. Vóila, what we have just obtained is a simple equation defining a *hyperbola*. Relativistic, uniformly accelerated motion takes place along a hyperbola, not a parabola as in the non-relativistic case. For short times, however, when the velocity of the rocket is still small, the hyperbolic trajectory can be expanded in a Taylor series and compared to the non-relativistic uniformly accelerated parabolic motion:

$$x(t) = \sqrt{K + c^2t^2} \approx \sqrt{K} + \frac{c^2t^2}{2\sqrt{K}} \equiv \text{const} + \frac{at^2}{2}, \tag{8.8}$$

where a is the proper acceleration of the rocket. Now we can connect the dots: $K = \frac{c^4}{a^2}$ and finally present the rocket's accelerated trajectory in its full glory:

$$x(t) = \frac{c^2}{a}\sqrt{1 + \frac{a^2t^2}{c^2}}. \tag{8.9}$$

The family of hyperbolas for different accelerations a, but sharing a common asymptote $ct = x$ is shown in Fig. 8.3. Notice that choosing a negative K would lead us to the uniformly accelerated trajectory of a superluminal body.

The same result can be re-derived using properties of four-acceleration (8.3). In an instantaneously co-moving reference frame the velocity of the rocket vanishes and the four-acceleration reduces

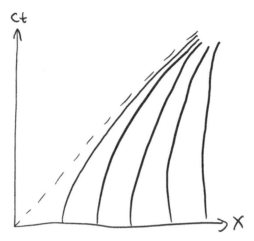

Figure 8.3:   Hyperbolic trajectories of a uniformly accelerated relativistic motion for various values of proper acceleration.

to $(0, a)$. But the square length of a four-vector is the same in all inertial frames, so we can compare the square length of the four-vector (8.3) with the square length of $(0, a)$:

$$-a^2 = -\frac{\left(\frac{\mathrm{d}v}{\mathrm{d}t}\right)^2}{(1 - v^2/c^2)^3}. \tag{8.10}$$

Therefore,

$$a = \frac{\frac{\mathrm{d}v}{\mathrm{d}t}}{(1 - v^2/c^2)^{3/2}} = \frac{\mathrm{d}}{\mathrm{d}t}\frac{v}{\sqrt{1 - v^2/c^2}}, \tag{8.11}$$

which can be solved with the initial condition $v(0) = 0$, yielding: $at = \frac{v}{\sqrt{1-v^2/c^2}}$. After a simple reversal we obtain

$$v(t) = \frac{at}{\sqrt{1 + a^2t^2/c^2}}, \tag{8.12}$$

and another integration over time brings us back to the familiar equation (8.9) defining a hyperbola. Notice that as t goes to infinity, the velocity in (8.12) tends to c, but never exceeds it, no matter how high the acceleration a is.

Our elegant result (8.9) holds an intriguing secret. Imagine a rocket that started to accelerate with some proper acceleration a

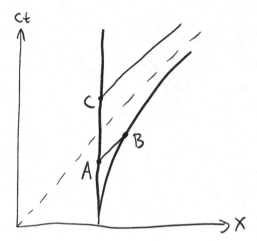

Figure 8.4: An accelerating rocket is being shot at with a laser gun located at the launch station. The first beam is shot at A and hits the rocket at B. The second beam is shot at C.

along a hyperbola. Another observer left behind at the launch station decided to destroy the escaping rocket with a laser cannon — see Fig. 8.4. At event A shown in the figure, that individual shoots the laser towards the rocket, the beam of which hits the rocket at B. Unfortunately, the rocket is not fully destroyed, so our persistent gunman aims and fires again at C. When will that shot hit the rocket?

The trajectory of the rocket approaches the asymptote $ct = x$, but never crosses it. For that reason, the second laser beam, which was shot beyond the asymptote, will never reach the rocket, even though it is moving with the speed of light! How is this possible? After all, the light should approach the accelerating observer with the constant speed c! Well, this example shows that the light does *not* do that at all. The fact light "always" moves with the same speed *only* applies to inertial observers, not accelerated ones. Soon, we'll discover more physical laws governing non-inertial frames of reference, but we have already established a very important one: the speed of light in an accelerated frame *can* differ from c.

It is interesting to realise that no information sent from beyond that asymptote will ever reach the rocket. No event that takes place *beyond* the asymptote will have *any* impact, or leave any physical trace on the accelerating observer. Even if an atomic bomb

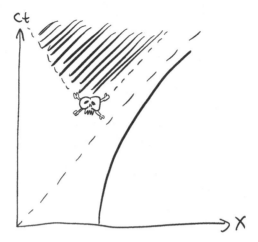

Figure 8.5: No event taking place beyond the event horizon, even the explosion of an atomic bomb, can affect the accelerated observer.

exploded, as shown in Fig. 8.5, the non-inertial observer inside the rocket would have no clue. The asymptote $ct = x$ determines the so-called *event horizon* of the accelerating observer and no information from the forbidden region of spacetime can ever cross it.

The event horizon is placed at the distance $\frac{c^2}{a}$ from the rocket, which can be read from equation (8.9). And that distance is the same for any inertial observer, instantaneously co-moving with the rocket. From now on, we will associate observations made by the accelerated observer in the rocket with the observations of the iner-tial, co-moving observer. Therefore for the uniformly accelerated individual, the event horizon is also placed at a fixed proper dis-tance of $\frac{c^2}{a}$. It's as if they were hovering above the event horizon of a static black hole. As we will soon discover, this analogy is not accidental.

8.3 Bell's Paradox

During one of his visits to CERN, John Bell (of *Bell's inequalities* fame), kept heckling the other physicists with a single yes-or-no question in the field of special relativity. Legend has it that they all gave him the *wrong* answer. Here is the riddle.

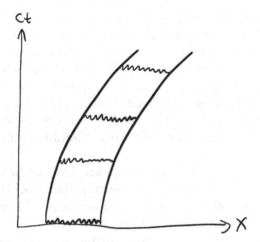

Figure 8.6: Two uniformly accelerated rockets separated by a fixed distance are connected by a tight rope.

Two resting rockets at distance L from each other are connected by a tight rope of the same length. In the same instant, both rockets start to move in the same direction, with identical uniform accelerations, while keeping their distance fixed — see Fig. 8.6. The question is simple: will the rope break or not? The question is worth digesting.

Here is the correct solution. The rope is moving with an increasing speed, therefore in the absence of any constraints it would normally shrink due to Lorentz contraction. However, both ends of the rope are attached to the rockets and therefore the rope is unable to contract. In fact, the rockets are stretching it to an unnatural length — its rest length. The stretching gradually increases and eventually breaks the rope. Sooner or later, the tail end of the rope will find itself beyond the event horizon of the front rocket; the binding forces of the rope will not be able to hold it together across the event horizon.

8.4 The Geometry of a Uniformly Accelerated Frame

We will now investigate the point of view of the non-inertial observers sitting inside the rockets. Our ultimate goal is to construct a uniformly accelerated frame of reference and understand its physics. But let us start with some basics. As we have already pointed out, we assume that the observations (of time and space) made by the accelerating non-inertial observer coincide with the observations made from an inertial, instantaneously co-moving inertial frame. This assumption includes the clock postulate stating that only the instantaneous velocity, not acceleration, can affect the rate of the moving clock. The same applies to measurements of space: the proper length of ideal rulers should not be affected by acceleration.

Let us first figure out, how do the accelerating observers interpret the breaking of the rope connecting their rockets? The only reasonable explanation is that, in their non-inertial frames, the distance between the rockets is actually increasing. Eventually, when the rope is unable to stretch any further — it snaps. Therefore, the distance between the rockets is fixed only in the inertial frame of reference. As a consequence, the rockets are not in a relative rest at all. The front rocket gradually moves away from the rear rocket.

This conclusion brings us to an interesting question. How should we modify the trajectory of the front rocket, so that the relative distance between the rockets is fixed in their own, non-inertial reference frame? In the previous example, the two rockets were moving away from each other, so the modified trajectory should lie somewhere in between the two hyperbolas from Bell's riddle. So, what is the unknown trajectory? The answer is simple and elegant: both rockets must accelerate along two hyperbolas sharing the common event horizon, as the ones shown in Fig. 8.3. Therefore, the front rocket has to move with a smaller proper acceleration than the rear rocket. The explanation is surprisingly simple. Each hyperbola lies at a fixed *proper distance* from the event horizon, which for a proper acceleration a is equal to $\frac{c^2}{a}$. Therefore all hyperbolas must also lie at fixed proper distances from each other!

We conclude that the family of hyperbolas sharing the common event horizon contains all the trajectories of accelerated observers that are in a relative rest. To an inertial observer these hyperbolas approach each other. However, to the accelerated observer their mutual distance is fixed. This has important implications for accelerated objects of finite size. Our discovery means that to accelerate such an object, each part of the body has to undergo a different proper acceleration; parts in the rear have to accelerate more than those in the front. Otherwise, the accelerated body would be torn to pieces.

The family of hyperbolas shown in Fig. 8.3 defines trajectories at a fixed proper distance from the event horizon. They characterise half of the uniformly accelerated *coordinate chart*. In order to complete that chart, we must define the *planes of simultaneity*. One such plane is obviously $ct = 0$, when all the accelerated rockets are instantaneously at rest. All the remaining planes of simultaneity can be obtained by Lorentz transforming the first one, which tilts them at an angle depending on the velocity. The resulting coordinate chart of the uniformly accelerated frame of reference is shown in Fig. 8.7. According to this chart, two arbitrary events lying on the same hyperbola take place at the same distance from the event horizon. Similarly, two arbitrary events lying on the same plane of simultaneity are simultaneous in the accelerated frame.

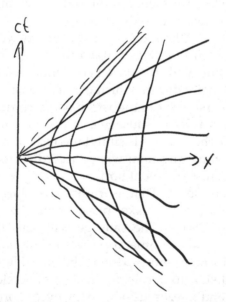

Figure 8.7:   Coordinate chart of the uniformly accelerated reference frame. Hyperbolas correspond to fixed positions, while tilted planes characterise the planes of simultaneity.

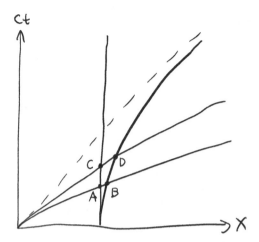

Figure 8.8:   A sausage fell out of the accelerating rocket and started to move freely. When the rocket was at B, the sausage was at A, when the rocket was at D, the sausage was at C according to the non-inertial observer.

Let us take a look at some of the intriguing consequences of this chart. Imagine a sausage that fell out of the uniformly accelerated rocket and started to move freely in space. This situation is depicted in Fig. 8.8. Without the loss of generality we have assumed that the sausage fell out at $t = 0$, so its free trajectory is just a straight vertical line while the rocket accelerates along a hyperbola. Planes of simultaneity dictate which events are simultaneous from the non-inertial perspective of the astronaut. According to this chart, when the rocket is at B, the sausage is at the same time at A. When the rocket is at D, the sausage is simultaneously at C, and so on. So, let us ask a puzzling question: when (according to the astronaut) will the sausage cross the event horizon?

The answer is *never!* That is right, according to our chart, that situation will never take place from the perspective of the accelerated observer. The sausage will start falling towards the event horizon, where time appears to slow down. The closer to the event horizon, the more extreme the time dilation. Eventually the sausage will start to decelerate instead of accelerating, until it comes to a stop and freezes just above the event horizon.

On the other hand, according to the inertial frame of the sausage, it will easily cross the horizon and move on without experiencing anything special about the crossing point. This shows that the physics of non-inertial frames pushes the relativity of events even further than special relativity did. Some events happening in the inertial frame will never take place for the non-inertial observer!

Let's play with some numbers and assume that the rocket accelerates with the proper acceleration g, so that the virtual force pushing down the accelerated observer feels like Earthly gravity. Sounds relatable. And let us assume that the sausage has an expiration date exactly one year from the day if fell from the rocket. How much of the astronaut's time will lapse before the sausage expires? This problem can be solved on the back of an envelope, as we will see in just a moment.

First of all, from (8.9) we conclude that the sausage fell out from the rocket at the distance $\frac{c^2}{g}$ from the event horizon, and the sausage's proper time remaining until it crosses the horizon equals $\frac{c}{g}$. Now, let us plug in numbers:

$$\frac{c}{g} \approx 3 \cdot 10^7 \ [\text{s}]. \tag{8.13}$$

But on the other hand,

$$1 \ \text{year} \approx \pi \cdot 10^7 \ [\text{s}]. \tag{8.14}$$

Therefore, $\frac{c}{g} < 1$ year, proves that the sausage will never expire, remaining fresh forever! Delicious!

There is even more intriguing stuff going on here. Have a look at Fig. 8.9. It can be seen that, although the sausage will eventually cross the event horizon (in its own reference frame) and lose the ability to send messages to the rocket, it can easily receive messages sent from the rocket. So, *one-way communication* through the event horizon *is* possible. This is exactly what happens close to a static *black hole*, therefore the asymptote $ct = x$ mimics the *black hole horizon*.

On the other hand, the other asymptote given by the equation $ct = -x$ has the exact opposite properties. Whatever happens on the other side of the horizon must affect the rocket, but nothing from the rocket can ever go back across the horizon. That other horizon has the properties of the so-called *white hole horizon*. The *white holes* are hypothetical objects that emit everything, but cannot absorb anything, just like time-reversed black holes. Notice that

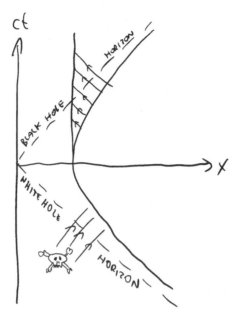

Figure 8.9:   Information can cross each of the event horizons only one-way. The accelerated observer can send any information across the black hole horizon, but will not receive a response. Any information can reach the rocket through a white hole horizon, but nothing can be sent back there by the non-inertial observer.

both of these horizons *overlap* at the same proper distance $\frac{c^2}{g}$ beneath the accelerated rocket.

8.5   Gravitational Time Dilation

Imagine a uniformly accelerated finite-size rocket carrying two ideal clocks at two different heights — see their hyperbolic trajectories in Fig. 8.10. As we already know, these two clocks will experience two different proper accelerations. Suppose that these clocks were synchronised at $t = 0$. Will they remain synchronised at a later plane of simultaneity of the accelerated frame depicted in Fig. 8.10?

The short answer is: no. It is clear from the figure that the path of clock 2 in front is longer than the path of clock 1 in the back, which means that the frontal clock will be ahead of the one at the

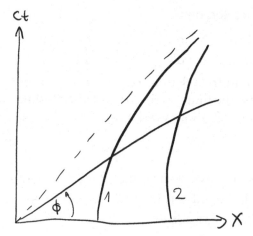

Figure 8.10: Two ideal clocks at relative rest are initially synchronised at $t = 0$. At a later plane of simultaneity tilted at an angle ϕ their reading is compared.

rear ruining the synchronisation. The striking conclusion is that time, at different heights in the accelerating rocket, runs at different rates.

We keep referring to these clocks as the one in the "front" or in the "back" of the rocket, while it should be more natural to label these clocks "higher" and "lower". After all, the accelerated rocket is not so unlike a rocket placed at rest in a gravitational field. We'll get back to that soon. Anyway, the effect that we have just discovered is called *gravitational time dilation*. The clock that is located lower (clock 1) will tick more slowly than the clock that is located higher (clock 2). The gravitational time dilation gets stronger as the clocks approach the event horizon, where time dilation becomes infinite and time stops.

Let's calculate how the time dilation depends on the distance from the event horizon. Take the two clocks from Fig. 8.10 and consider the next plane of simultaneity at an infinitesimally small angle $d\phi$. In that case, the velocities of both clocks vanish and the proper time lapsed along each path will simply be equal to the length of that path, which is proportional to the product of the angle $d\phi$ and the distance from the event horizon. As a consequence, the ratio of the proper times at the two chosen paths will be equal to the ratio of distances from the event horizon, which are inversely proportional to the proper accelerations:

$$\frac{d\tau_1}{d\tau_2} = \frac{c^2/a_1}{c^2/a_2} = \frac{a_2}{a_1}. \tag{8.15}$$

In order to obtain the relation between finite proper times, we can simply compose the finite angle ϕ out of infinitesimal angles $d\phi$ and iterate the above equation, yielding:

$$\frac{\tau_1}{\tau_2} = \frac{a_2}{a_1}. \tag{8.16}$$

If the lower clock 1 experiences the proper acceleration $a_1 = g$ and the difference in heights of the clocks is h, then $\frac{c^2}{a_2} = \frac{c^2}{g} + h$, which combined with (8.16) leads to:

$$\tau_2 = \tau_1 \left(1 + \frac{gh}{c^2}\right). \tag{8.17}$$

Notice that for $h = -\frac{c^2}{g}$, i.e. at the event horizon, time comes to a complete stop. This is why, according to the accelerated observer, nothing can cross the horizon. If we ever take a trip in an accelerating rocket, it makes sense to store all the food in the lower part of the rocket where it will remain fresh for longer. And one more observation: if the front of the rocket moves with acceleration a then the whole rocket cannot be longer than $\frac{c^2}{a}$.

The formula (8.17) characterises the clock rates at different heights in the accelerating rocket. In Section 8.11, we will justify why this also applies to clocks hanging in the gravitational field. It follows that our head ages a little bit faster than our feet, because they are a little closer to Earth.

8.6 Rindler Transformation

This course on relativity began with the derivation of the Lorentz transformation characterising a transition between two inertial observers. Now that we have familiarised ourselves with the geometry of the uniformly accelerated frame called the *Rindler frame*, it is time to characterise the transition between an inertial and a uniformly accelerated observer, known as the *Rindler transformation*.

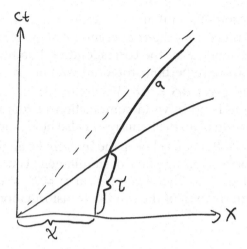

Figure 8.11: Rindler hyperbolas are numbered by their proper distance from the event horizon, and planes of simultaneity are labelled by the proper time measured by the reference clock at their intersection.

The coordinate chart shown in Fig. 8.7 needs one more detail in order to become a coordinate system. We must enumerate all the hyperbolic trajectories of constant position, as well as the planes of simultaneity. The hyperbolas will be enumerated with their proper distance χ from the event horizon. Therefore, a hyperbola characterised by the proper acceleration a will be numbered with $\chi \equiv \frac{c^2}{a}$ in the Rindler frame — see Fig. 8.11.

In order to enumerate the planes of simultaneity we first pick a reference hyperbola with a chosen proper acceleration a. Then the time assigned to a given plane of simultaneity is simply the proper time τ of the clock moving along the reference hyperbola (measured from $t = 0$) at the intersection with that plane — see Fig. 8.11.

Any event (ct, x) within the domain of the Rindler frame, $x > |ct|$, lies at an intersection of some hyperbola χ with some plane of simultaneity τ — see Fig. 8.12. Our present goal is to determine the relation between the coordinates assigned to that event by the inertial (ct, x) and non-inertial $(c\tau, \chi)$ observer. Let us first rewrite equation (8.9) in the form: $x^2 - c^2 t^2 = \frac{c^4}{a^2}$. Using the association $\chi = \frac{c^2}{a}$ we obtain the first transformation equation: $\chi = \sqrt{x^2 - c^2 t^2}$.

In order to derive the second equation we use the previously established relation between the infinitesimal proper time and the angle of the plane of simultaneity $c\mathrm{d}\tau = \frac{c^2}{a}\mathrm{d}\phi$. In Section 1.2 of Chapter 1, we noted that the trigonometric angle ϕ can be related to a hyperbolic angle ϑ of the corresponding Lorentz transformation (which is just a hyperbolic rotation) via $\tan\phi = \tanh\vartheta$, which for small angles gives $\mathrm{d}\phi = \mathrm{d}\vartheta$. Therefore $c\mathrm{d}\tau = \frac{c^2}{a}\mathrm{d}\vartheta$. Since the hyperbolic angle is additive (a composition of rotations equals a rotation by the sum of angles), this result also holds for finite values. Therefore, $c\tau = \frac{c^2}{a}\vartheta$. Let us choose the hyperbolic angle ϑ in such a way that the corresponding plane of simultaneity contains the event (ct, x) — see Fig. 8.12. Then we have $\tan\phi = \frac{ct}{x} = \tanh\vartheta$, which completes the derivation of the Rindler transformation:

$$c\tau = \frac{c^2}{a}\operatorname{atanh}\frac{ct}{x},$$

$$\chi = \sqrt{x^2 - c^2 t^2}.$$
(8.18)

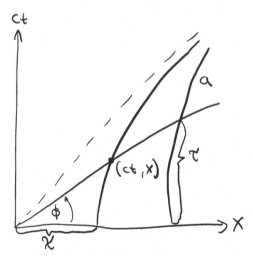

Figure 8.12: Every event (ct, x) lies on the intersection between some hyperbola and some plane of simultaneity. The corresponding Rindler coordinates $(c\tau, \chi)$ of that event are such that χ is the distance of that hyperbola from the event horizon and τ is the proper time of the reference clock timed at the intersection with that plane of simultaneity.

Perpendicular coordinates are transformed trivially: $y' = y$ and $z' = z$. The inverse Rindler transformation to (8.18) has a more elegant form:

$$ct = \chi \, \sinh\left(\frac{a\tau}{c}\right),$$
$$x = \chi \, \cosh\left(\frac{a\tau}{c}\right). \tag{8.19}$$

Rindler transformation is only valid in the quarter of spacetime that is accessible to the uniformly accelerated observer, and given by $x > |ct|$. We should remember that the parameter a appearing in the equations is not the actual acceleration of the observer, as the Rindler chart covers the full range of proper accelerations from zero to infinity. The parameter a functions as a convention to enumerate the planes of simultaneity and does not have to be physically significant in any way. For small accelerating objects it is convenient to match the reference acceleration with the proper acceleration of such an object so that the proper time and coordinate time also match.

Finally, let us find out how the spacetime interval changes under Rindler transformation. From (8.19) we obtain

$$c^2 dt^2 - dx^2 - dy^2 - dz^2 = \frac{a^2 \chi^2}{c^4} c^2 d\tau^2 - d\chi^2 - dy'^2 - dz'^2. \tag{8.20}$$

As we can see the spacetime interval expressed in terms of Rindler coordinates takes a new, modified form, which signifies a physical difference between the family of inertial observers and the uniformly accelerated observer.

8.7 Free Motion According to the Rindler Observer

Let us return to the story of the sausage falling from the rocket moving with acceleration g, as shown in Fig. 8.8. Let us investigate how the aging process of the sausage proceeds according to the non-inertial observer accelerating in the rocket. The task turns out to be simple; all we have to do is take the first of equations (8.18), substitute $a = g$ to use the rocket's ideal clock as the reference time and plug in the position of the sausage, which is $x = \frac{c^2}{g}$. We obtain

$$t = \frac{c}{g} \tanh\left(\frac{g\tau}{c}\right). \tag{8.21}$$

It is that simple! As we see, the sausage will never expire, because $t < \frac{c}{g} < 1$ year. The sausage's aging will gradually slow down as it nears the event horizon, and will asymptotically will come to a stop just above the abyss.

Another question worth asking is what is the trajectory of the sausage in the rocket's frame? Substituting $x = \frac{c^2}{g}$ and $a = g$ into the second equation (8.19), we get

$$\chi = \frac{c^2/g}{\cosh\left(g\tau/c\right)}. \tag{8.22}$$

Again we see that even if we wait forever, the sausage will never cross the event horizon. We can even determine how the sausage's velocity changes according to the observations of the astronaut:

$$v_c \equiv \frac{d\chi}{d\tau} = -c\frac{\sinh\left(\frac{g\tau}{c}\right)}{\cosh^2\left(\frac{g\tau}{c}\right)}. \tag{8.23}$$

This velocity is the so-called *coordinate velocity*, and we plot it against time in Fig. 8.13. We can see that initially the sausage speeds up, but after the time $\frac{c}{g}\mathrm{asinh}(1) \approx \frac{c}{g}$ the velocity reaches the maximum value of $\frac{c}{2}$ and then begins to decrease due to the increased gravitational time dilation. Eventually, the sausage stops just above the event horizon. Let us briefly remind ourselves that "at the same time" nothing particularly interesting takes place in sausage's inertial reference frame. It just easily, unremarkably easily crosses the event horizon without noticing anything to write home about.

And what about light? Can it win against the freezing powers of the event horizon? Consider a beam of light shot from the rocket towards the event horizon, as shown in Fig. 8.14. This ray is moving along the trajectory: $ct = -x + \frac{c^2}{g}$. In order to determine its path in the accelerated frame of reference, we will substitute both equations

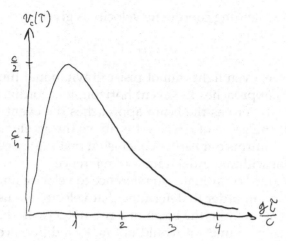

Figure 8.13: Coordinate velocity of the freely moving sausage according to the accelerated astronaut, as a function of his proper time. After a time of roughly $\frac{c}{g}$, the velocity reaches the maximum of $\frac{c}{2}$.

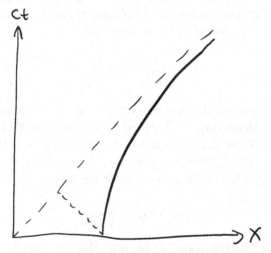

Figure 8.14: A light beam shot from an accelerating rocket towards the event horizon slows down and stops according to the observer inside the rocket.

(8.19) with $a = g$ into the above equation which, after the simplification, gives us

$$\chi = \frac{c^2}{g} e^{-\frac{g\tau}{c}} \tag{8.24}$$

and the corresponding coordinate velocity is given by

$$v_c = ce^{-\frac{g\tau}{c}}. \tag{8.25}$$

As we can see, even light cannot resist gravitational time dilation. As the beam approaches the event horizon, its coordinate velocity slows down to zero as the beam approaches the event horizon. It also exceeds the value of c when it is above the accelerated rocket, which only confirms our earlier conclusion that the speed of light is only constant within inertial frames of reference.

The adjective "coordinate" in reference to velocity underlines an important fact. In order to determine that velocity we used a clock that was far away from the beam of light, not a clock the light was passing. Such a distinction would not make a difference in *inertial* frames, where all the clocks are ticking at the same rate. It *does* make a difference, however, for an accelerated observer. If we asked "what would the velocity measured by the accelerated local clock be at the moment in which the moving object passes it?", the result would be completely different. A velocity measured by a local clock is called the *local velocity* and it is given by

$$v \equiv \frac{d\chi}{d\tau_{\text{loc}}} = v_c \frac{c^2}{g\chi}, \tag{8.26}$$

where we have used the relation $d\tau_{\text{loc}} = d\tau \frac{g\chi}{c^2}$ between local time $d\tau_{\text{loc}}$ and reference time τ obtained from equation (8.15). For light the local velocity is always equal to c, which can be seen by substituting (8.24) and (8.25) into (8.26). For the sausage that is moving according to (8.23), the local velocity (8.26) reads:

$$v = -c \tanh\left(\frac{g\tau}{c}\right) \tag{8.27}$$

and it approaches the speed of light as the sausage approaches the event horizon.

The notion of local velocity is meaningful both physically and operationally, unlike the coordinate velocity, which is rather misleading. Making measurements using distant devices like clocks or rulers that are necessary for the definition of coordinate velocity makes little physical sense; we should keep in mind that it is based on the notion of simultaneity, which is rather sloppy. Just recall the

Elvis Presley paradox discussed in Section 2.8. On this basis, we should be careful when making declarations about where an object that has just fallen from an accelerating rocket is "now". The next example should aid our understanding.

8.8 Hungry, Hungry Astronauts

We return to the odyssey of the falling sausage which, as we know, never crosses the event horizon from the rocket-bound astronaut's perspective, no matter how much time passes inside the rocket. But does this mean the sausage is still accessible and could be brought back on board at any time? For instance, could the astronaut send out a rescue mission to grab the sausage and return it to the rocket? Only for a while.

How about we take a gander at Fig. 8.15, where the whole situation is shown? We can see from that figure that the last moment when the astronaut can still hope to recover the sausage takes place at $\Delta\tau$ of his proper time after it fell from the rocket. A simple calculation shows that if the rocket's proper acceleration was equal to g, then $\Delta\tau = \frac{c}{g} \log 2$. After that time no subluminal mission will be able to reach the sausage before it falls beyond the event horizon —

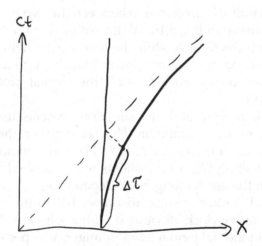

Figure 8.15: A sausage has fallen from an accelerating rocket. The non-inertial observer has time $\Delta\tau$ to send out a rescue mission to bring the sausage back onboard. After that time, it will be too late.

and our astronaut will remain hungry. So, we could claim that the sausage "has not yet" fallen beyond the event horizon, but what does it really mean if there is no chance to feed the astronaut? Also, whatever signal the astronaut tries to send the sausage after time $\Delta\tau$ has passed, it will reach the sausage already on the other side of the event horizon. Hence, no reply to that message will ever be received by the rocket. The last ray of hope for the astronaut to taste the sausage ever again is if the sausage itself decides to return to the rocket of its own volition by turning its own engine on. So, to summarise, the question of which side of the horizon the sausage is on is not only operationally ambiguous, but more importantly, not so meaningful.

8.9   The Twin Paradox According to the Accelerated Twin

It is typical to dismiss the twin paradox discussed in Section 2.6 by stating that the twin accelerated in his-rocket does not belong to the family of inertial observers, therefore his observations are essentially worthless.

Well, we are now in a position to fully investigate the accelerated twin's point of view. All we need to do is assume that his rocket is moving with uniform acceleration between the two meeting points, A and B, as shown in Fig. 8.16. We'll assume that the brothers synchronised their clocks at A while the non-inertial twin was already moving with a non-zero initial velocity. Then he started decelerating and eventually turned back to meet his inertial brother at B to compare clocks.

First, we'll analyse how the situation presents itself from the point of view of the inertial twin. This scenario is rather trivial. We only have to substitute the velocity (8.12) into equation (2.11) and vóila! The resulting Fig. 8.17(a) shows with a dashed line how the time lapse on the accelerating clock depends on the time indicated by the inertial clock. For comparison we left in the uniform time lapse of the inertial clock, illustrated with a solid line. No surprises there. The non-inertial twin returns younger, as expected.

Now, let's consider the situation from the point of view of the accelerated astronaut twin. It is easiest to use the first formula

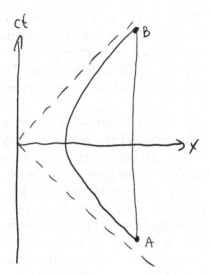

Figure 8.16: A twin paradox scenario with one of the twins in a uniformly accelerated motion between the two meeting points, A and B.

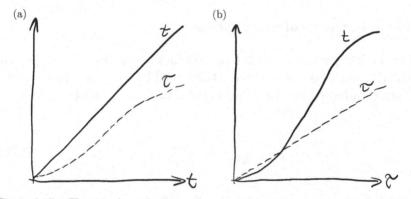

Figure 8.17: The time lapse between the meeting points as witnessed by (a) the inertial twin, (b) the non-inertial twin, as a function of their own proper times. The solid line corresponds to the ticking of the resting clock, while the dashed line characterises the time lapse on the accelerated clock.

from the Rindler transformation (8.18). We need only ensure that both clocks are synchronised at the initial point A. Using elementary operations on our equations we can obtain the plot illustrating the ticking of both clocks as a function of the proper time of the accelerated clock. The results are shown in Fig. 8.17(b). As we can see, at first the astronaut can observe the kinematic time dilation of his inertial brother's clock. But after a while, when the inertial clock reaches greater heights, its rate increases. Because of this, the inertial twin becomes older and stays this way until both brothers meet at **B**.

By comparing plot 8.17(a) with 8.17(b), we see that the conclusions drawn in both frames of reference are fully consistent — in both plots, the final age gap between the twins is identical.

Finally, let's make a challenging observation. If the astronaut twin changed his mind half-way through the journey and decided to accelerate away from his twin at $t = 0$, it is possible that his inertial brother began moving *backward* in time and becoming younger. Such a scenario could happen if the inertial twin ended up beyond the temporary event horizon created by the accelerated motion of his astronaut brother. This is yet another consequence of having too much trust in the concept of the plane of simultaneity.

8.10   The Energy of a Free Particle

Let us first point out an interesting fact about the motion of our favorite space bound sausage, as observed by the accelerating astronaut. Combining our two results (8.22) and (8.27) yields:

$$\sqrt{1 - \frac{v^2}{c^2}} = \frac{g\chi}{c^2}, \tag{8.28}$$

which leads to the conclusion that the following quantity is constant in that free motion:

$$E = \frac{mg\chi}{\sqrt{1 - v^2/c^2}}, \tag{8.29}$$

where the mass of the moving object m has been added for reasons that will soon become clear. When the velocity is non-relativistic, $v \ll c$ and the distance from the event horizon does not change much, i.e. $\chi = \frac{c^2}{g} + h$, where h is small, we have

$$E \approx mc^2 + mgh + \frac{mv^2}{2} + \cdots. \tag{8.30}$$

What are we looking at? Is this some sort of conserved energy in free motion? As we have already pointed out, physics in a uniformly accelerated frame is very similar to physics in a static gravitational field with gravity replaced by non-inertial virtual forces. The similarities extend to the presence of the energy conservation law. A freely moving sausage is like a freely falling object due to gravity, where the sum of kinetic and potential energies is constant. However, our non-approximated expression (8.29) does not involve the sum of energies — it is just a single, mononomial expression. We are so used to adding new terms to the "total" energy that we do not often ask, why should the energy be additive? Well, apparently it isn't! Keep in mind that *Noether's theorem* only states that time translation invariance results in the conservation of energy, that is all. We've already witnessed how rest energy is inseparable from kinetic energy in special relativity. Now we can see that the relativistic "potential" energy of the virtual force is inextricably merged with the others. This dynamic will repeat when we study conservation of energy in a real, relativistic static gravitational field.

To close this section, let's point out that the newly discovered conserved energy (8.29) depends on the local velocity v, not its coordinate counterpart v_c. This supports our earlier claim that only locally measurable quantities are physically significant.

8.11 The Principle of Equivalence

Long ago, we stepped beyond the domain of orthodox special relativity which is the study of properties of the inertial frames. We will now take another big step and realise that gravity is fundamentally a relativistic phenomenon.

One thing that makes gravity absolutely unique is its universality. Gravity affects all objects exactly the same way, which may not sound very surprising, but just you wait! Soon we will realise that this inconspicuous property is of great significance. Imagine a room with no windows and a little switch on the wall. This switch can turn gravity on and off. We'll start with it in the off position. Now, let's pick a handful of objects: a chair, a pen, a feather, a helium filled balloon, an electron, a positron, and whatever else we desire, and place them levitating at rest in various positions. If we flip the switch, all the objects will fall, but the balloon will fly up. But why up? Because the air inside the room also falls down, pushing the lighter helium balloon upwards. So, what if we suck all the air out of the room and repeat the experiment? This time all the objects will fall, and they will fall with identical accelerations. This is an observable fact that has no known counterexamples: all objects known to humans fall in a gravitational field the exact same way. And here comes the fun part! There is one more way to make all the objects fall in exactly the same way, but without switching on the gravity. Guess what it is?

The answer is surprisingly simple. If, instead of gravity the switch initiates a rocket engine hidden underneath the floor that makes the entire room accelerate upwards, the effect will be exactly the same. All the levitating objects will be approached by the floor, which is indistinguishable from the scenario in which they all fell down. This observation is the reason for the famous principle of equivalence: in a sufficiently small volume, the consequences of gravity are indistinguishable from the consequences of non-inertial motion of a whole reference frame. This equivalence does not hold globally for all of spacetime, but it works locally within any selected small region.

We can now see why universality of gravity is so crucial. If even one exotic object was found that moved differently in a gravitational field, classical or quantum, made of anti-matter or anything else, it would immediately invalidate the principle of equivalence. Moreover, if this principle is true, even light should fall in the gravitational field. Not only fall, but fall with the exact same acceleration as everything else. Indeed, all the photons surrounding us right now must freely fall with acceleration g. If there are any doubts, let's take a second look at equation (8.25):

$$v_c = ce^{-\frac{g\tau}{c}} \approx c - g\tau. \tag{8.31}$$

The photons move so fast that their fall is hard to observe, but precise optical experiments give us perfect confirmation of this fact.

We can now realise why our study of the accelerated Rindler frame was so important. All the peculiar effects involving gravitational time dilation, black hole and white hole event horizons should be expected from a *real* gravitational field! Our intuitive picture of the accelerated astronaut we developed in this chapter can therefore serve as a perfect playground to help us understand all the most relevant aspects of the physics of black holes before we even look at the equations of general relativity! We should now understand, why it does not seem possible for an outside observer to throw anything into a black hole, whereas it is perfectly possible for a free-falling object to cross the event horizon. What happens beyond? We shall discover that soon.

It is still extremely puzzling that gravity is completely indistinguishable from a virtual force due to acceleration of the observer. Or, shall we say that gravity *is* just a virtual force? If this is true, then the search for the quantum version of gravity would not differ much from the search for the quantised virtual force that is acting on us whenever we take a bus ride and the driver makes a sudden turn or hits the brakes.

8.12   Questions

- What is relativistic uniformly accelerated motion?
- What force can generate uniformly accelerated motion? And how does this force vary in time in the accelerated frame?
- Is it true that an object that starts uniformly accelerating at a given point can always be caught up with by a light beam released from the same point some time later?
- How can a rigid body of a finite size be released from an accelerating rocket in such a way that the body can continue to move freely without any internal forces present?
- What happens to a ray of light sent towards the event horizon from an accelerating rocket? And what if the light was shot in the opposite direction?
- What is the difference between coordinate velocity and local velocity?
- Can the coordinate velocity of a massive body exceed the speed of light?

• The uniformly accelerated frame is a stationary construct with no preferred moment of time, but different points in space (height) do differ. What causes this asymmetry between time and space?

8.13   Exercises

• Derive the inverse Rindler transformation (8.19) from the Rindler transformation (8.18).

• Derive the result mentioned in Section 8.8 showing how much time the accelerating astronaut has to make a decision about recovering his sausage.

• Suppose a rescue mission to save the freely falling sausage is dispatched at a different height inside the rocket than the one at which the sausage was released. Which initial height offers the most (proper) time to make a decision about launching the rescue mission?

• Derive a non-relativistic formula for the conserved energy of an object of mass m freely moving in a non-relativistic, uniformly accelerated frame.

• An accelerated observer sends a light signal towards a mirror placed at rest at the top of the rocket. After the proper time $\Delta\tau$, he receives the reflected signal back. Calculate the height of the mirror. Is it given by the standard radar formula $\Delta\chi = \frac{c\Delta\tau}{2}$? Based on the result, what can we deduce about the coordinate and local speed of light in the non-inertial frame?

• An accelerated observer sends a light signal of frequency v towards the event horizon. Calculate the dependence of the frequency of light on the distance from the event horizon at which the light was registered.

• Determine the equation of the future light cone of any event observed in the uniformly accelerated frame.

• Derive the trajectory of a body in oblique projectile motion within the uniformly accelerated relativistic rocket.

• Determine spacetime surfaces at which all ideal clocks moving along the family of accelerated hyperbolas remain synchronised. Why do we not use these surfaces to define the planes of simultaneity of the accelerated frame?

Chapter 9

Curved Spacetimes*

9.1 Spacetime Metric

What is fascinating about non-inertial frames is not just the virtual forces that modify the physics within such frames, but also the event horizons, variable time rate at different heights, and other phenomena we are about to discover. The principle of equivalence prepared us to expect similar effects in gravity, which will be explored in this Chapter. Tumbling down this rabbit hole will eventually take us across an event horizon towards the very heart of a static black hole and white hole. But let us start off with a fun surprise.

Consider two reference frames — a resting, inertial frame K and a non-inertial frame K' that rotates around the $z = z'$ axis with the angular velocity ω. Let these two observers examine a circle centred symmetrically around the rotation axis, as shown in Fig. 9.1. Firstly, the inertial observer in K measures the circumference of that circle and its diameter using a set of short, stiff rulers that she places along the measured shape to tightly fill in its total length. By counting how many rulers are needed to cover the whole circumference, and how many of them to cover the diameter, she finds that the ratio of these numbers is equal to π. But we've been warned a surprise is coming, so let's test the circle in the rotating frame K'. The measurements of

---

*This is a more advanced chapter. In order to follow all the details a knowledge of elementary differential equations, Euler–Lagrange equations and the Gauss theorem is needed.

Figure 9.1:   A rotating observer measures the circumference and the diameter of the circle drawn on the ground.

the diameter in this frame yield the exact same outcome, because all the rulers are moving *perpendicular* to their length. But the measurement of the circumference will be affected by the Lorentz contraction, because all of the rulers are moving *along* their length and hence become shorter by the Lorentz factor $\sqrt{1 - \omega^2 r^2/c^2}$, where r is the distance from the z' axis. As a consequence, more rulers will fit along the circle yielding a longer circumference in K' and, as a result, its ratio to the diameter will be greater than π. The new formula for the circumference of a circle centred around the z' axis takes the form:

$$\text{(circumference)} = \frac{2\pi r}{\sqrt{1 - \omega^2 r^2/c^2}}. \tag{9.1}$$

Farewell to Euclidean geometry! Space in the non-inertial frame K' is *non-Euclidean*; the sum of interior angles of a triangle is *not* equal to π. Therefore, in order to determine distances between points in such geometry a non-trivial *space metric* needs to be employed.

There are other interesting phenomena that affect the time rate in the rotating frame K'. A clock that is resting in K' will tick slower due to time dilation if it is further away from the z' axis compared to a clock placed in the centre. The time dilation in the non-inertial frame increases with the distance from the centre as a result of kinematic time dilation (as witnessed by the inertial observer). It is the exact opposite of gravitational time dilation on Earth, wherein the effect grows stronger as we approach the centre. The reason for this discrepancy is that the gravitational force is directed inwards, while the virtual centrifugal force of the rotating frame is pointing outwards.

It should not come as a surprise that non-Euclidean effects also take place in the presence of gravitational fields. In general, space-time can become *curved* due to gravity. To characterise its properties, we will introduce the notion of *spacetime metric* known and often employed in topology. As we know, the distance between space-time events is measured by their spacetime interval. In all inertial frames it has an invariant, simple form. However, for non-inertial observers that form becomes more complicated, as we have already witnessed in expression (8.20). In general, a spacetime interval in an arbitrary, non-inertial frame has the form:

$$ds^2 = g_{\mu\nu}dx^\mu dx^\nu, \tag{9.2}$$

where dx^μ is the μ component of the derivative of the four-position and $g_{\mu\nu}$ is the *spacetime metric* (or the so-called *metric tensor*) characterising the physics of the non-inertial frame. It can be way different from the Minkowski metric $\eta_{\mu\nu}$ given by (7.3) that characterises inertial frames. For example, the spacetime metric in the Rindler reference frame accelerating along the direction x that is expressed in (8.20) can be written as

$$g_{\mu\nu} = \begin{pmatrix} a^2\chi^2/c^4 & 0 & 0 & 0 \\ 0 & -1 & 0 & 0 \\ 0 & 0 & -1 & 0 \\ 0 & 0 & 0 & -1 \end{pmatrix}. \tag{9.3}$$

How complicated can the metric be in general? The corresponding 4×4 matrix has 16 elements, but since it has to be symmetric (the anti-symmetric component wouldn't contribute to interval ds^2 anyway): $g_{\mu\nu} = g_{\nu\mu}$, only 10 of these elements are independent. Also, the whole spacetime can be characterised with infinitely many coordinate systems. The 3D space can be described with Cartesian, spherical, cylindrical coordinate systems, as well as and many others. A coordinate transformation between these systems affects the metric and, therefore, can be used to make some of its elements vanish. Four coordinate transformation equations allow us to eliminate four elements of the metric leaving us with only six independent elements. In general, they can all depend on the spacetime coordinates x^μ.

Let's investigate how the individual elements of the matrix relate to measurements of time and space. Let us denote the coordinates used by a given non-inertial observer by (x^0, x^1, x^2, x^3). Keep in mind that these coordinates do not strictly correspond to the "real" time and "real" length, just like the angle ϕ of the cylindrical chart does not correspond to any particular "length". These coordinates only serve to unambiguously enumerate all the events in spacetime.

So, how to calculate the actual time rate $d\tau$ of an ideal clock placed at rest at a given point in space? Since our clock is not moving, we have $dx^1 = dx^2 = dx^3 = 0$. And for any given metric $g_{\mu\nu}$ the proper time between infinitesimally close events, is in this case proportional to the spacetime interval between these events:

$$d\tau = \frac{1}{c}\sqrt{g_{00}}dx^0. \tag{9.4}$$

As we can see, the above expression only makes sense for $g_{00} > 0$. This condition must be met if we want our clocks to make any physical sense and measure real time.

Let us move on and investigate what are the proper distances dl measured between events in the non-inertial frames. For simplicity, we will limit ourselves only to a scenario in which the following metric elements vanish: $g_{01} = g_{02} = g_{03} = 0$. Fortunately it is always possible to make them vanish via a simple coordinate transformation. In this case the study becomes much simpler, and an arbitrary pair of simultaneous events is separated by the proper distance dl given by

$$dl^2 = -g_{ij}dx^i dx^j, \tag{9.5}$$

where Roman indices i, j, as opposed to Greek ones μ, ν, cover the set containing only values $1, 2$, and 3.

Our results allow us to get a grip on the physical interpretation of the metric elements g_{00} and g_{ij}, as quantities characterising the local lapse of proper time, and local proper distances.

We can also derive the expression for the local velocity of an object observed in a non-inertial frame. By combining the formula

(9.4) with (9.5) we obtain

$$v^2 = \left(\frac{\mathrm{d}l}{\mathrm{d}\tau}\right)^2 = -c^2 \frac{g_{ij}\mathrm{d}x^i\mathrm{d}x^j}{g_{00}(\mathrm{d}x^0)^2}. \tag{9.6}$$

All the standard expressions for special-relativistic time dilation (2.2) and Lorentz contraction (2.3) are still valid as long as the local velocity (9.6) is substituted. Finally, let us invoke the principle of equivalence and declare that all the results discussed here are also valid in real gravitational fields.

9.2 A Free Fall in Curved Spacetime

In Section 8.10 of Chapter 8, we have found the expression (8.29) for the conserved energy of a free body moving in the Rindler frame. Energy is also conserved in all non-inertial frames with a static metric, i.e. such that $g_{\mu\nu}$ does not depend on the temporal coordinate x^0. The same applies to gravitational fields: if the field is static then all freely moving test bodies will conserve their energy. We will now derive the expression for that energy using two methods: a rough, hand-waving argument, followed by a more sophisticated, strict calculation to confirm that result.

When the non-inertial motion of the observer is 1D, like for the uniformly accelerated Rindler frame, then the metric elements do not depend on the perpendicular coordinates y and z. In this case, the thickness of freely moving objects measured in these perpendicular directions is constant in time and the equations of motion along y and z are trivial.

Static metrics that do not depend on the temporal coordinate x^0 have a similar property. Imagine an iPod playing a song. The playing coordinate time Δx^0 will be the same regardless of the position in which the iPod was placed. We will use this fact for the derivation of the expression for the energy. Suppose that a freely falling iPod that plays an infinitely short sound lasting $\mathrm{d}x^0$ of the coordinate time. The sound is played twice, exactly when the iPod passes by two ideal clocks placed at r_1 and r_2 — see Fig. 9.2. The coordinate time must be equal for both sounds, therefore, using (9.4)

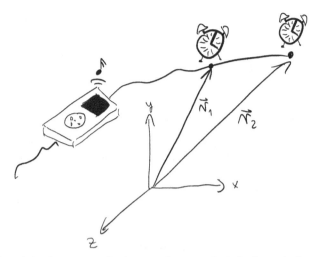

Figure 9.2:   A freely moving iPod passes by two ideal clocks and plays two identical short sounds.

we find:

$$\frac{d\tau_1}{\sqrt{g_{00}(r_1)}} = \frac{d\tau_2}{\sqrt{g_{00}(r_2)}}, \tag{9.7}$$

where $d\tau_i$ are the playback times measured by both ideal clocks. Since each of the measured times, $d\tau_i$, depends on the instantaneous local velocity v_i of the iPod relative to the clocks, given by (9.6), we have

$$\frac{d\tau\sqrt{1 - v_1^2/c^2}}{\sqrt{g_{00}(r_1)}} = \frac{d\tau\sqrt{1 - v_2^2/c^2}}{\sqrt{g_{00}(r_2)}}, \tag{9.8}$$

where $d\tau$ is the playback proper time (measured in the inertial frame instantaneously co-moving with the iPod). The obtained result allows us to define the conserved energy of any massive body freely moving in a static spacetime (whether of a non-inertial frame or gravitational field):

$$E \equiv \frac{mc^2\sqrt{g_{00}(r)}}{\sqrt{1 - v^2(r)/c^2}}, \tag{9.9}$$

where r is the position of the body and v is its instantaneous local velocity. The multiplicative term mc^2 was added so that the obtained expression could reduce to the standard formula for the energy of a free body (7.16) when $g_{00} = 1$.

So far, we have assumed the existence of instantaneously co-moving inertial frames that we can always jump into. Such a global reference frame can only exist if the spacetime is flat, which is not the case when gravity is involved. For curved spacetimes the notion of instantaneously co-moving inertial frames is limited to sufficiently small regions in which such constructs can be introduced. Therefore, from now on, whenever we refer to such a frame it is implied that this frame only exists locally. In practice, such a frame is one that is freely falling in a given gravitational field, within a given small region.

Let us now derive the formula (9.9) in a rigorous way. In order to do so, we will formulate a general principle governing the free motion of test bodies moving freely in a given gravitational field. So, what is so special about free motion? In inertial frames of reference the free motion is the one that takes place along a straight line, which defines the shortest *distance* between any pair of points. Therefore, one could employ the *variational principle* that minimises a certain *Lagrangian* to define such a motion. Unfortunately, such a Lagrangian would not be relativistically invariant, because the length is frame dependent.

Let us find another characteristic of the free motion that has a relativistic structure. In Section 2.6 where we discussed the twin paradox, we established that the astronaut twin travelling in the accelerated rocket ends up younger than his inertial brother. It is clear from equation (2.11) that this is the case for any non-inertial trajectory of the rocket. This shows that the free (inertial) trajectory connecting two points corresponds to the longest possible *proper time* among all possible trajectories. This enlightening observation suggests another possible definition of free motion, that involves the variational principle. Since the proper time $d\tau$ is proportional to the spacetime interval ds, we can introduce the following *principle of least action*, leading to the trajectory for which the proper time is maximum:

$$\int \mathcal{L} dx^0 \equiv -mc \int ds = \text{(minimum)}, \qquad (9.10)$$

where $\mathcal{L}$ is the *Lagrangian*. The multiplicative constant involving the mass m of the moving body was included for a reason that will soon be clear. As we can see, the trajectory of the body is completely characterised by the spacetime metric. The above definition also applies when the spacetime is curved due to gravity. This is because, for any local patch of spacetime, the principle of equivalence states that the situation is indistinguishable from a scenario involving a non-inertial frame. Therefore, each patch of the trajectory minimises the action (9.10). And since that action is additive, the global action is also minimised.

The principle of least action (9.10) applies to general dynamical spacetimes, including time dependent ones. But let us consider a special case of that principle: when the spacetime involved is static (i.e. it does not depend on the temporal coordinate x^0) and $g_{0i} = 0$. In this case the Lagrangian $\mathcal{L} = -mc^2 \frac{ds}{dx^0}$ is time-independent and by taking $ds = \sqrt{g_{00} dx_0^2 + g_{ij} x^i x^j}$ reduces to

$$\mathcal{L} = -mc^2 \sqrt{g_{00} + g_{ij} \dot{x}^i \dot{x}^j / c^2}, \qquad (9.11)$$

where $\dot{x}^i \equiv c \frac{dx^i}{dx^0}$, $i \in \{1, 2, 3\}$. For time-independent Lagrangians there exists the following constant of motion E, known as the energy:

$$
\begin{aligned}
E &\equiv \dot{x}^i \frac{\partial \mathcal{L}}{\partial \dot{x}^i} - \mathcal{L} = -mc^2 \frac{g_{ij} \dot{x}^i \dot{x}^j / c^2}{\sqrt{g_{00} + g_{ij} \dot{x}^i \dot{x}^j / c^2}} + mc^2 \sqrt{g_{00} + g_{ij} \dot{x}^i \dot{x}^j / c^2} \\
&= \frac{mc^2 g_{00}}{\sqrt{g_{00} + g_{ij} \dot{x}^i \dot{x}^j / c^2}} = \frac{mc^2 \sqrt{g_{00}}}{\sqrt{1 - v^2/c^2}}, \qquad (9.12)
\end{aligned}
$$

where we have used the definition of the local velocity (9.6). We can see that our result agrees with the previously derived formula (9.9). As we have already discussed, the resulting expression contains the rest energy, kinetic energy and gravitational potential energy combined. The Lagrange formalism also shows that, for time-dependent metrics resulting in time-dependent Lagrangians, energy is not conserved in free motion.

9.3 Maximum Damage Induced by a Stinking Egg

Here is a simple problem for us to think about. As we know, a chicken is a device produced by an egg, for the purpose of producing other eggs. Suppose we want to pick an egg and throw it at someone. The chicken at our disposal produces a suitable egg at a given time and place — event **A**. Our target shows up at a distant location, at some later time, for just a moment — event **B**. But being hit by an egg is not humiliating enough! To maximise the damage and suffering, we would love the egg to be old and stinking. So, what could we do for the egg to become as old and stinking as possible? We can attach it to a drone and steer it between A and B in such a way that the proper time is maximised the moment the egg hits its target. How could we determine this optimum trajectory and turn our egg into the most terrifying biological weapon?

On one hand, we do not want the egg to move too fast because time dilation would reduce the aging process. Therefore it seems best that the egg moves along the shortest possible path so that the velocity can be minimised — see the trajectory 1 in Fig. 9.3. On the other hand, that short path is located close to the Earth, where the gravitational time dilation is strongest. In order to avoid it, the egg should move as far from Earth and as high as possible — see the trajectory 3 in Fig. 9.3. But then the velocity would also have to be high so that the egg reaches B on time... It seems that the optimum

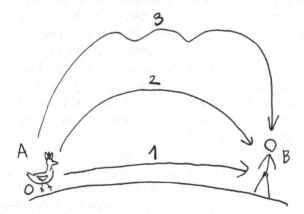

Figure 9.3: Using a biological weapon against an unsuspecting target. Which trajectory will lead to the maximum aging of the flying egg?

choice must lie somewhere in between these paths, around trajectory 2 in Fig. 9.3. But how can the optimum choice be determined?

The task may seem difficult, but it is not. So, how should we steer our drone? We shouldn't! We should just throw the egg such that it lands in a given place at a given time, at the event B, and turn the drone's engines off. The reason that this is the correct choice follows directly from the principle of least action introduced in (9.10). According to that principle, among all possible trajectories the one that defines free motion involves the maximum proper time between its ends. Therefore, the gravitational field already *knows* how to solve our dilemma, and does it for us!

9.4   Black Hole

So far, we have considered properties of a given spacetime metric without asking where that particular metric comes from. In order to provide a comprehensive answer, we'd need to know how to solve Einstein's equations of general relativity. These complicated, non-linear equations are rather hard to deal with. So, instead of going down this route, which would take us beyond the scope of this book, we will take a shortcut and try to guess a few things.

Notice that one lesson from our previous considerations is that the virtual forces present in a uniformly accelerated frame are not uniform in space. The proper acceleration that is experienced in the accelerating rocket decreases with height. The higher we are in the rocket, the weaker is the virtual force we experience. Due to the principle of equivalence the same can be stated about gravity. No uniform gravitational field is possible; the gravitational force must decrease as the distance from its source increases. How exactly does gravitational force change with height? This is one of the questions we will figure out in this section.

We will consider one of the simplest, and yet physically most interesting, cases and attack the problem of the spacetime metric of a static *black hole* of mass M. It will not be a strict derivation; we will need some hand-waving to justify some of the steps. However, it will provide us with valuable intuitions behind the physics of black holes.

Earth and its gravity is like a ball and chain to us. In both New-tonian physics and general relativity, the gravitational pull from a massive point-like object is indistinguishable from the one created by a finite ball of the same mass. When no angular momentum is involved a spherically symmetric mass produces a spherically symmetric spacetime metric with time-independent elements. In conve-nient to use spherical coordinates in which the point-like mass M is placed at the origin $r = 0$ the metric takes the general form:

$$ds^2 = g_{00}(r)c^2dt^2 + g_{rr}(r)dr^2 - r^2d\Omega^2, \qquad (9.13)$$

where $g_{00}(r)$ and $g_{rr}(r)$ are some unknown functions and $d\Omega^2 \equiv d\theta^2 + \sin^2\theta d\phi^2$ is the angular element of the spherical coordinate system. Here, the radial coordinate r has been dialled in such a way that it is equal to the corresponding circle circumference centred at the origin of the coordinate system, divided by 2π (or alternatively, so that its square, r^2, is equal to the surface of a sphere, divided by 4π). The actual proper length along the radial direction is given by $\sqrt{-g_{rr}(r)}dr$ in accordance with equation (9.5). We expect and assume that far away from the mass (i.e. in the limit of $r \to \infty$) the spacetime becomes Euclidean which means that the unknown functions must satisfy $g_{00}(\infty) = 1$ and $g_{rr}(\infty) = -1$.

In order to determine $g_{00}(r)$ and $g_{rr}(r)$ we will consider the fol-lowing thought experiment based on the original idea presented in [21]. Imagine a set of n identical cages of a shape similar to a trape-zoid, as shown in Fig. 9.4. The top segment of each cage is composed of a circle arc of the radius $r + dr$ and length $\frac{2\pi(r+dr)}{n}$, while the bot-tom is a circle arc of a radius r and the length $\frac{2\pi r}{n}$. The remaining straight sides have the length dr, as shown in Fig. 9.4. Let us place all the n cages in an empty space without any gravity involved, along a large circle with the bottom segments towards the centre and push them gently inwards as shown in Fig. 9.4. We will assume that the radial velocity is so insignificantly small that Lorentz con-traction can be neglected. The dimensions of the cages were chosen in such a way that they can simultaneously bunch together to form a tight ring of thickness dr — see Fig. 9.4. In the inertial frame of one of these cages, the collision between the neighbouring walls of the connecting cages will take place at a single instant.

Now suppose that an identical experiment with our n cages is carried out in the gravitational field characterised by the metric

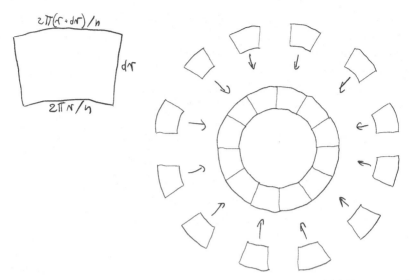

Figure 9.4: A thought experiment with n identical cages freely falling towards the centre.

(9.13). Again, we initially position all the cages along a large circle centred around $r = 0$ (sufficiently large so that initially the gravity is negligibly small), and give them a tiny push towards the centre. According to the principle of equivalence, the observations carried out inside one of these cages should be indistinguishable from the results of the previous experiment. Therefore, once more the cages' side walls of length dr have to connect at a single instant with their neighbours.[a] This time, however, the collision takes place in the region of finite r, where geometry is affected by gravity and the proper thickness of the resulting ring of the circumference equal to $2\pi r$ is not dr, but $\sqrt{-g_{rr}(r)}dr$. Also, now the moving cages have some non-negligible velocity that gives rise to the Lorentz contraction. These two effects have to cancel each other out in order to avoid conflict with the principle of equivalence. Therefore, the height of the moving cages just before the collision must

---

[a]This argument may be questioned, because we consider two separated regions within two neighbouring cages, while the principle of equivalence only applies locally. However, our reasoning leads to the correct result, so we will carelessly ignore this concern.

remain dr:

$$dr = \sqrt{1 - \frac{v^2(r)}{c^2}}\sqrt{-g_{rr}(r)}dr, \qquad (9.14)$$

where $v(r)$ is the velocity of the cages moments before collision.

The velocity $v(r)$ can be determined using the conservation of energy (9.9) condition. The initial resting position of the cages is far away from the black hole, therefore,

$$mc^2 = \frac{mc^2\sqrt{g_{00}(r)}}{\sqrt{1 - \frac{v^2(r)}{c^2}}}, \qquad (9.15)$$

where m is the mass of a single cage that cancels out anyway. Comparing equations (9.14) and (9.15) leads to the first important conclusion:

$$g_{00}(r)g_{rr}(r) = -1. \qquad (9.16)$$

Let us return to the physics within the Rindler frame uniformly accelerated along the direction x. The acceleration $g(r)$ of a freely falling object placed at a given point r of that frame is inversely proportional to the distance from the event horizon, x:

$$g(r) = -\frac{c^2}{x}\hat{x}. \qquad (9.17)$$

Such an acceleration is perceived, in the non-inertial frame, as the effect of the virtual forces present in that frame. The field of accelerations $g(r)$ of the freely falling test bodies satisfies an interesting condition, which can be found by calculating the divergence of the above equation:

$$\nabla \cdot g(r) = \frac{c^2}{x^2} = \frac{g^2(r)}{c^2}. \qquad (9.18)$$

This nonlinear relation shows that the virtual forces within the non-inertial frame are not additive (unlike the analogous electric forces acting on test charges placed at an arbitrary point, for which the divergence vanishes in empty space). As a consequence, the virtual forces present within the non-inertial frames cannot satisfy linear equations. Indeed, Einstein's equations of general relativity are highly nonlinear.

Invoking the principle of equivalence, we will assume that the same condition (9.18) is satisfied for the gravitational field of the black hole, since the physics in these two scenarios should be locally indistinguishable. Let us apply formula (9.18) to the case of the "real" gravitational field. Consider an infinitesimally thin, spherical shell with the surface $4\pi r^2$ (corresponding to the radial coordinate r), that has the proper thickness equal to $\sqrt{-g_{rr}(r)}\mathrm{d}r$. We will apply the Gauss theorem to transform the divergence of the vector field appearing in equation (9.18) to the integral form involving the stream of that field through the boundary surface. In our case, the acceleration field $\mathbf{g}(\mathbf{r})$ is radially oriented inwards, and its magnitude $g(r)$ depends only on the magnitude r of the vector $\mathbf{r}$. Equation (9.18) transformed into the Gaussian form for the corresponding volume element reads:

$$-g(r+\mathrm{d}r)4\pi(r+\mathrm{d}r)^2 + g(r)4\pi r^2 = \frac{g^2(r)}{c^2}4\pi r^2\sqrt{-g_{rr}(r)}\mathrm{d}r, \quad (9.19)$$

which can be written as

$$\frac{\mathrm{d}\left(g(r)r^2\right)}{\mathrm{d}r} = -\frac{g^2(r)}{c^2}r^2\sqrt{-g_{rr}(r)}, \quad (9.20)$$

and, by using (9.16), it can be cast in the form:

$$-\frac{\sqrt{g_{00}(r)}}{g(r)r^2}\frac{\mathrm{d}\left(g(r)r^2\right)}{\mathrm{d}r} = \frac{g(r)}{c^2}. \quad (9.21)$$

This is our second important result, which will be used shortly.

We need one more independent equation on top of (9.16) and (9.21) in order to determine the three unknown, but related, functions: $g_{00}(r)$, $g_{rr}(r)$, and $g(r)$. Let us consider the infinitely short free fall of a test mass m from the initial radial position $r + \mathrm{d}r$ to the final position r. Since the developed velocity is infinitesimally small, we can approximate the energy (9.9) in the final point as

$$E = \frac{mc^2\sqrt{g_{00}}}{\sqrt{1 - v^2/c^2}} \approx mc^2\sqrt{g_{00}} + \frac{mv^2}{2}\sqrt{g_{00}}, \quad (9.22)$$

and use the non-relativistic expression for the energy conservation:
$\frac{mv^2}{2} = mg\sqrt{-g_{rr}}\mathrm{d}r$ to write the energy conservation law in the following form:

$$mc^2\sqrt{g_{00}(r+\mathrm{d}r)} = mc^2\sqrt{g_{00}(r)} + mg(r)\sqrt{-g_{rr}(r)}\sqrt{g_{00}(r)}\mathrm{d}r,$$
(9.23)

which combined with (9.16) gives us the third important equation:

$$\frac{\mathrm{d}\sqrt{g_{00}(r)}}{\mathrm{d}r} = \frac{g(r)}{c^2}.$$
(9.24)

The remaining part is easy — we have derived a complete set of three simple differential equations (9.16), (9.21), (9.24), and we need to solve them. By comparing the left-hand sides of equations (9.21) and (9.24) we obtain

$$\frac{\mathrm{d}\left(\sqrt{g_{00}(r)}g(r)r^2\right)}{\mathrm{d}r} = 0,$$
(9.25)

which has a trivial solution equal to some constant C:

$$\sqrt{g_{00}(r)}g(r)r^2 = C.$$
(9.26)

By substituting $g(r)$ from equation (9.24) we obtain

$$\frac{\mathrm{d}g_{00}(r)}{\mathrm{d}r} = \frac{2C}{c^2r^2},$$
(9.27)

which can be directly integrated with the boundary condition $g_{00}(\infty) = 1$, yielding:

$$g_{00}(r) = 1 - \frac{2C}{c^2r},$$

$$g_{rr}(r) = -\frac{1}{1 - 2C/c^2r},$$
(9.28)

$$g(r) = \frac{C}{r^2\sqrt{1 - 2C/c^2r}}.$$

The missing constant C can be determined in the non-relativistic limit $\lim_{c\to\infty} g(r) = \frac{C}{r^2}$ using the approximate *Newton law of universal gravitation* $g(r) \approx \frac{GM}{r^2}$, where G is the *Newton constant,* yielding $C =$ GM. Therefore, the final form of the spacetime metric of the static black hole, famously known as the *Schwarzschild metric* takes the form:

$$ds^2 = \left(1 - \frac{2GM}{c^2 r}\right) c^2 dt^2 - \left(1 - \frac{2GM}{c^2 r}\right)^{-1} dr^2 - r^2 d\Omega^2, \quad (9.29)$$

with the resulting field of accelerations:

$$g(r) = \frac{GM}{r^2 \sqrt{1 - 2GM/c^2 r}}. \quad (9.30)$$

The obtained expressions for the metric (9.29) and the field of accelerations (9.30) misbehave at the radial coordinate distance $r = \frac{2GM}{c^2} \equiv R$, known as the *Schwarzschild radius,* and at the point $r = 0$ known as the *singularity.* Notice that the local velocity of a freely falling body of a finite energy E approaching the Schwarzschild radius must tend to the speed of light, because the numerator of the expression (9.9) tends to zero. Therefore no subluminal test object may escape from beneath the surface $r = R$ known as the *event horizon.*

The sign change of metric elements at the event horizon makes it appear as though time and space interchanged their roles. This behaviour reflects the fact that the coordinates we have used to characterise the spacetime of a static black hole correspond to a family of resting observers placed around the black hole. Such observers can exist outside of the event horizon, but in order for them to exist beyond the event horizon, they have to be superluminal. Notice that the sign flip in the metric elements signifies the transition from a subluminal to a superluminal family of stationary observers residing at fixed distances from the singularity. As we have discovered in Section 4.1, and later in the formulas (4.14), such a transition involves an interchange between time and space, exactly as we are witnessing here. This also shows that, although it is possible to rule out superluminal observers from special relativity by adding an extra "impossibility" postulate (in spite of the fact that they are both mathematically and physically allowed), these observers inevitably return when we get to in general relativity. Without them, the physical interpretation of the Schwarzschild metric (9.29) can only cause headaches.

9.5 Testing the Principle of Equivalence

Let's investigate the properties of the black hole spacetime in the proximity of the event horizon, $r \approx R$. We first introduce the new variable $\Delta r = r - R$, so that the Schwarzschild metric (9.29) takes the form:

$$ds^2 = \frac{\Delta r}{\Delta r + R} c^2 dt^2 - \frac{\Delta r + R}{\Delta r} d\Delta r^2 - r^2 d\Omega^2. \qquad (9.31)$$

Near the Schwarzschild radius, $r \approx R$, i.e. for $\Delta r \ll R$, the infinitesimal proper distance (9.5) can be approximated by

$$dl = \sqrt{-g_{rr}(r)} dr \approx \sqrt{\frac{R}{\Delta r}} d\Delta r = 2d\sqrt{R\Delta r}, \qquad (9.32)$$

which can be integrated, giving $l = 2\sqrt{R\Delta r}$ and substituted into (9.31), yielding the approximate form of the metric near the event horizon:

$$ds^2 = \frac{l^2}{4R^2} c^2 dt^2 - dl^2 - r^2 d\Omega^2. \qquad (9.33)$$

And so an exciting result has just emerged: we have obtained the familiar metric of the uniformly accelerated Rindler observer (9.3) with $\chi \equiv l$ and $a \equiv \frac{c^2}{2R}$. This shows that the laws of physics in the proximity of the black hole event horizon are indeed indistinguishable from the physics laws near the Rindler event horizon of the accelerating rocket. As we have already argued, no subluminal object, nor even light, can escape from beneath the horizon. Everything we have already learned in Chapter 8 about the reality of uniformly accelerated frames applies directly to the description of black holes.

We should also mention one more mysterious detail. In Fig. 8.10, describing the Rindler frame, we noticed the presence of two event horizons: a black hole horizon and a white hole horizon. The latter is transmitting all the information across towards the outside observer and not absorbing anything. We should note that the spacetime (9.29) corresponds not only to a black hole, but to a black hole and a white hole combined. This is because the obtained metric is an idealisation, while the real-life black holes dwelling in our universe are typically collapsing stars, that become (9.29) only in the asymptotic limit of the infinite future.

## 9.6	Falling Under the Event Horizon

From the point of view of the outside observer watching a sausage fall into the static black hole, the sausage's story never crosses the horizon. From their perspective the sausage just freezes above. So, if we adapt this view, nothing ever falls into the black hole. We have to recall, of course, the limitations of the concept of the plane of simultaneity that we are using here. We have already discussed its drawbacks in Section 2.8.

But how about the point of view of another observer freely falling with the sausage? We know that, for such an observer, crossing the event horizon is only a matter of time, and nothing remarkable happens at that moment (at least in the absence of the quantum effects). But what happens afterwards? Can the sausage reach the singularity at $r = 0$ within finite proper time of its own? We will not be able to answer this question by only investigating the properties of the Rindler frame. We have to deal with the full spacetime metric (9.29) directly. Let us consider the simplest case of the radial free fall into the black hole, by taking $d\Omega = 0$, in which case the spacetime interval element (9.29) along the freely falling trajectory reduces to

$$ds^2 = \left(1 - \frac{2GM}{c^2 r}\right)c^2 dt^2 - \left(1 - \frac{2GM}{c^2 r}\right)^{-1} dr^2. \tag{9.34}$$

The definition of the local velocity (9.6) leads to the following relation:

$$g_{00}(r)dt^2 = -\frac{g_{rr}(r)}{v^2}dr^2, \tag{9.35}$$

which, upon substitution into (9.34), yields:

$$ds^2 = \frac{1}{1 - 2GM/c^2 r}\left(\frac{c^2}{v^2} - 1\right)dr^2. \tag{9.36}$$

We can calculate the local velocity v of the freely falling body at position r by employing the law of the conservation of energy (9.9). Let us assume that the resting body was released from initial position r_0. Conservation of energy leads to the equation:

$$\frac{c^2}{v^2} - 1 = \frac{1 - 2GM/c^2 r}{2GM/c^2 r - 2GM/c^2 r_0}, \tag{9.37}$$

that we can substitute into (9.36) getting rid of the velocity dependence. Since the spacetime interval is proportional to the proper time of the falling body, $ds = c d\tau$, we obtain

$$d\tau^2 = \frac{dr^2}{2GM\,(1/r - 1/r_0)}. \tag{9.38}$$

All we have left to do is integrate the above expression between r_0 and 0. It is easiest to parametrise the trajectory in the following way: $r(\eta) = \frac{r_0}{2}(1 + \cos\eta)$, which leads to

$$d\tau = \sqrt{\frac{r_0^3}{8GM}}(1 + \cos\eta)d\eta. \tag{9.39}$$

Integrating the above expression is straightforward and results in the following proper time that lasts until the freely falling object reaches the singularity:

$$\tau = \pi\sqrt{\frac{r_0^3}{8GM}}. \tag{9.40}$$

This proper time is finite. If we start at the event horizon, the expression reduces to: $\tau = \pi\frac{GM}{c^3}$. That is how much time the free-falling observer has left before they reach the unknown. This result completes our introduction to the topic of black holes and curved spacetimes. The most fascinating aspects of black holes are still left out until we learn about *quantum theory*, but we'll save those for another occasion.

9.7 Questions

- What is the difference between the metric in a 3D space and that of a 4D spacetime?
- Is it always possible to introduce a coordinate system in which the metric of a curved spacetime reduces to the Minkowski metric?
- Is energy always conserved for a freely falling object in an arbitrary spacetime?
- Why is the trajectory of free motion a solution to the variational problem?
- What is the difference between the event horizon of a static black hole and the event horizon of a uniformly accelerated observer?

- What happens to the information falling into the black hole?
- According to the outside observer, it is not possible for any object to reach the event horizon within a finite time. Is it possible for a star to gravitationally collapse giving rise to the creation of a black hole, within finite time?
- What would happen to the trajectory of the Earth if the Sun suddenly collapsed into a black hole?
- What is the local speed of light under an event horizon?
- Consider a stationary observer hovering at a fixed close distance from the event horizon of a static black hole and watching a freely moving object. In this case his observations can be described to a good approximation using the Rindler accelerated observer as shown in Fig. 9.5. The freely moving object seems to emerge right from the event horizon — does it contradict the claim that nothing can escape from beyond that horizon?

9.8   Exercises

- Use the radar method to determine the expression for the proper length between two infinitely close events for a spacetime metric with $g_{0i} \neq 0$.

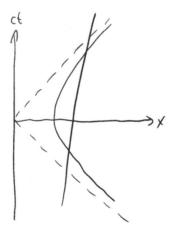

Figure 9.5:   A stationary observer close to the event horizon observing a freely moving object.

- What is the variational principle leading to the trajectory of a light ray in a curved spacetime?
- Compute the real surface area of the event horizon of a black hole.
- What is the Schwarzschild radius of a black hole of the same mass as the Earth? What is the error we make by assuming that the ratio of the length of the equator to the Earth's diameter is equal to π?
- Calculate the proper time along the free trajectory of a body freely falling from the distance r_0 to the event horizon of a black hole.
- Consider the interpretation of the sign change of the Schwarzschild metric beyond the event horizon involving the presence of superluminal static observers, for which time and space exchange their roles. Calculate the proper distance between the singularity of a black hole and its event horizon.

Chapter 10

Relativity of Electrodynamics*

10.1 Prelude

At long last, we have finished applying relativistic doctrines to kinematics, dynamics, and gravity. The time has come to take a closer look at relativistic aspects of the theory of electromagnetism. Maxwell's equations contain two phenomenological constants: vacuum permittivity ε_0 and vacuum permeability μ_0:

$$\nabla \cdot E = \frac{\varrho}{\varepsilon_0},$$

$$\nabla \cdot B = 0,$$

$$\nabla \times E = -\frac{\partial B}{\partial t}, \tag{10.1}$$

$$\nabla \times B = \mu_0 j + \mu_0 \varepsilon_0 \frac{\partial E}{\partial t}.$$

It turns out that the product of these two constants has the same physical unit as the inverse of the square velocity $[\frac{s^2}{m^2}]$. We can therefore introduce the new constant $c \equiv \frac{1}{\sqrt{\mu_0 \varepsilon_0}}$ and determine its value experimentally. This we will do by measuring the repulsive force of two electric charges and the attractive force of two electric currents,

---

*This is a more advanced chapter aimed for students with basic knowledge of Maxwell's equations and the mathematical properties of the ∇ differential operator.

and multiplying the resulting electromagnetic constants ε_0 and μ_0, respectively: $c \approx 3 \cdot 10^8 [\frac{m}{s}]$. What is the physical meaning of that constant? To figure it out we'll need to make a few simple transformations of equations (10.1). By taking a rotation of both sides of the third equation and using the first and the fourth equation in vacuum we obtain

$$\frac{1}{c^2} \frac{\partial^2 E}{\partial t^2} - \Delta E = 0. \tag{10.2}$$

The result is nothing more than a simple wave equation with the solutions propagating with the speed c. These solutions involve electromagnetic waves such as light, but also any other disturbances of the field, hence the name of the new constant: *the speed of light*. Does it already start to smell relativistic?

10.2   Electromagnetic Derivation of the Lorentz Transformation

We began this course by deriving the Lorentz transformation using a Hermann Minkowski method based on the assumption of the constancy of the speed of light. Later on, in Section 4.1, we developed an alternative approach that did not make use of that assumption, being based instead on the Galilean principle of relativity. Now, we would like to derive the same Lorentz transformation one last time. But this time we will base our derivation on electromagnetic considerations. Let's start off with an interesting example of the scalar electric potential of a moving point-like charged particle. Seems straightforward to derive, but we will soon realise how easy it is to make a mistake.

First, let us consider a general, time-dependent charge distribution $\varrho(r, t)$ localised within a certain volume V, and its effect on the scalar potential at a given point r:

$$\varphi(r, t) = \int_V d^3s \, \frac{\varrho(s, t)}{4\pi\varepsilon_0 |r - s|}. \tag{10.3}$$

To obtain this formula, we considered an infinitesimaly small charge $\varrho(s, t)d^3s$, wrote down the resulting Coulomb potential at r, and integrated over all charges. Unfortunately our formula is incorrect, because the scalar potential at a given instant t cannot depend on the distant charge distribution at that same instant! The potential φ simply cannot "know" about the instantaneous charge distribution. It only "knows" about the distribution at earlier times, otherwise the scalar potential φ could mediate instantaneous signalling. We need not refer to special relativity in order to justify our claim. According to equation (10.2), disturbances of the field can only propagate with the speed of light c. Therefore, the correct expression characterising the scalar potential from the charge distribution $\varrho(r, t)$ must be:

$$\varphi(r, t) = \int_V d^3s \, \frac{\varrho(s, t_{\text{ret}})}{4\pi\varepsilon_0 |r - s|}, \tag{10.4}$$

where $t_{\text{ret}} \equiv t - \frac{|r-s|}{c}$ is the so-called *retarded time*.

So, let us return to the earlier question: what is the scalar potential generated by a point-like charge q moving with a velocity v? The simple answer would involve a retarded Coulomb potential:

$$\varphi(r, t) = \frac{q}{4\pi\varepsilon_0 R_{\text{ret}}}, \tag{10.5}$$

where R_{ret} is the distance between the observation point and the position of the moving charge, $|s - r|$, at a retarded time. Unfortunately, we are wrong again, as the correct formula has an extra term:

$$\varphi(r, t) = \frac{q}{4\pi\varepsilon_0 R_{\text{ret}}} \frac{1}{(1 - v_r/c)_{\text{ret}}}. \tag{10.6}$$

That term depends on the retarded radial velocity component v_r along a direction pointing outward from the observation point r.

Where does the additional term come from? To understand its origin, let us consider a rod AC, uniformly charged with the total charge q, that is moving along its length [22] as shown at the space-time diagram in Fig. 10.1. Let's determine the scalar potential along the trajectory of the rod at a certain point B. The grey area of the figure represents the spacetime region occupied by the charged rod. The dashed line represents the information about the charge reaching the observer at B. It is seen that the apparent length of the rod given by the intersection of the dashed line with the grey area is

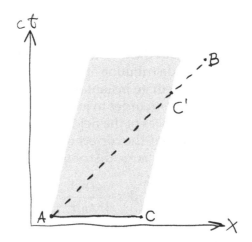

Figure 10.1: Spacetime diagram of a uniformly charged moving rod. The grey area represents the effective charge distribution smeared in space.

not $|AC|$, but $|AC'|$. Therefore, the apparent electric charge observed at B is not q, but $q\frac{|AC'|}{|AC|}$. To find that apparent charge we first determine the relation $|AC| = |AC'|(1 - v/c)$ following from elementary kinematic considerations. As a consequence, the apparent charge is equal to $\frac{q}{1-v/c}$ and does not depend on the length of the rod, therefore our result also holds for point-like charges. In a general, 3D case the velocity v is replaced with its radial component v_r relative to the observer. This justifies the extra term present in the correct formula (10.6).

Finally, we can rewrite the expression (10.6) in a more elegant form:

$$\varphi(r, t) = \frac{q}{4\pi\varepsilon_0 (R - R \cdot v/c)_{\text{ret}}}, \qquad (10.7)$$

where $R \equiv s - r$, and derive a similar expression for the vector potential A of a moving charge:

$$A(r, t) = \frac{q v_{\text{ret}}}{4\pi\varepsilon_0 (R - R \cdot v/c)_{\text{ret}}}. \qquad (10.8)$$

The resulting equations are known as the *Liénard–Wiechert potentials*. In order to derive the Lorentz transformation as promised, we will determine the Liénard–Wiechert potentials for the point-like charge q moving with a constant velocity v along x.

The task is rather straightforward. First, the relation between the retarded distance R_{ret} and the retarded time t_{ret} is given by

$$t_{\text{ret}} = t - \frac{R_{\text{ret}}}{c}. \tag{10.9}$$

Second, the equation of motion of the charge is simply $x = vt$, therefore the following relation is also satisfied:

$$R_{\text{ret}} = \sqrt{(x - vt_{\text{ret}})^2 + y^2 + z^2}. \tag{10.10}$$

Combining equations (10.9) and (10.10) we obtain the following quadratic equation for t_{ret}:

$$c^2(t - t_{\text{ret}})^2 = (x - vt_{\text{ret}})^2 + y^2 + z^2. \tag{10.11}$$

By choosing the one solution satisfying $t_{\text{ret}} < t$ we obtain

$$\left(1 - \frac{v^2}{c^2}\right) t_{\text{ret}} = t - \frac{vx}{c^2} - \frac{1}{c}\sqrt{(x - vt)^2 + \left(1 - \frac{v^2}{c^2}\right)(y^2 + z^2)}. \tag{10.12}$$

The retarded distance R_{ret} can be computed by combining the results (10.9) with (10.12). The scalar potential φ, given by (10.7), involves the retarded radial velocity, v_r, given by

$$(v_r)_{\text{ret}} = v\frac{x - vt_{\text{ret}}}{R_{\text{ret}}}, \tag{10.13}$$

therefore, the retarded denominator of expressions (10.7) and (10.8) reads:

$$\left(R - \frac{\boldsymbol{R} \cdot \boldsymbol{v}}{c}\right)_{\text{ret}} = c(t - t_{\text{ret}}) - \frac{v}{c}(x - vt_{\text{ret}}) = c\left[t - \frac{vx}{c^2} - \left(1 - \frac{v^2}{c^2}\right)t_{\text{ret}}\right]. \tag{10.14}$$

Substituting (10.12) into the above equation and then the result into (10.7), we finally obtain

$$\varphi(\boldsymbol{r}, t) = \frac{q}{4\pi\varepsilon_0} \frac{1}{\sqrt{1 - v^2/c^2}} \left[\left(\frac{x - vt}{\sqrt{1 - v^2/c^2}}\right)^2 + y^2 + z^2\right]^{-1/2}. \tag{10.15}$$

So far, we have only played with elementary properties of the Maxwell's equations (10.1) to determine the scalar potential of a uniformly moving point-like charge. The result revealed something unexpected — a familiar expression lurking within the resulting formula (10.15)! The derivation of the Lorentz transformation (1.7) is now straightforward. We only need to assume the principle of relativity, which states that the same result (10.15) should also be obtained by considering a resting charge and the observer moving with a uniform velocity in the opposite direction. This requirement imposes that the Lorentz transformation (1.7) is the only coordinate transformation that allows this. It transforms the scalar potential of the resting charge:

$$\varphi(r, t) = \frac{q}{4\pi\varepsilon_0} \frac{1}{\sqrt{x^2 + y^2 + z^2}} \tag{10.16}$$

into (10.15) given in the moving frame. The first transformation formula (1.7) can be directly extracted from the expression (10.15). Then we can write the inverse transformation (1.8) by simply flipping the sign of velocity, and by combining these two equations we can obtain the second transformation equation of (1.7). A similar path to the discovery of the Lorentz transformation was taken by Lorentz and Poincaré many years before Einstein.

It is sometimes claimed that relativity is already imprinted in Maxwell's equations. This is only partially true. Indeed, we do not have to modify Maxwell's equations to account for relativistic effects. Equations (10.1) are already correct. But the knowledge of the equations of electromagnetism is not sufficient to derive special relativity. The missing ingredient is the principle of relativity stating that Maxwell's equations must have the exact same form in all inertial reference frames. Without this principle, the Lorentz transformations cannot follow. And make no mistake, the application of the principle of relativity to the laws of electromagnetism was not obvious. It was more natural to think about electromagnetic fields as disturbances of some substance that had a preferred rest frame. Just like sound waves that propagate in the air also have a preferred reference frame.

From Section 4.1, we already know that the principle of relativity alone is enough to derive equations of special relativity, so for that Maxwell's equations are not really necessary. Historically speaking, though, it was electromagnetic equations and their symmetries that inspired the birth of the Lorentz transformation and the physics behind it. One thing we learn, though is that the mysterious constant K that appears in the transformation (4.7) is the inverse square of the speed of the actual light.

One more detail. The difference between equations (10.15) and (10.16) does not only boil down to coordinate transformation. There is an extra term $\sqrt{1 - v^2/c^2}$ in the denominator, whose origin will become fully clear later. Right now, we can only say that without that term the total charge would change between frames. This is not allowed for the reasons discussed in Section 2.5. This is because the integral of the Laplacian of the scalar potential φ would be different for a resting charge and for a moving charge. The extra term in the denominator prevents that from happening and guarantees that the total charge is the same in every inertial frame.

We have derived the scalar potential of a moving point-like charge. An analogous procedure for the vector potential A leads to the formula:

$$A = \frac{v}{c^2}\varphi. \tag{10.17}$$

We will soon discover a hidden connection between the potentials φ and A and a beautiful structure behind it.

10.3 A Relativistic Formulation of Electrodynamics

As we gradually unravel the relativistic structure behind the laws of electromagnetism, we still have many surprises ahead of us. Maxwell's equations are local differential equations and their solutions should transform relativistically. But what about the transformation properties of the equations themselves? Let us have a closer look at the differential operators appearing in (10.1).

A change of reference frame involves a coordinate transformation from (ct, r) into (ct', r'). For the observer moving with a uniform velocity V along x, the corresponding Lorentz transformation (1.7) affects the differential operators $\frac{\partial}{c\,\partial t}$ and $\nabla \equiv (\nabla^x, \nabla^y, \nabla^x)$ in the following way:

$$\frac{\partial}{c\,\partial t'} = \frac{\partial t}{\partial t'}\frac{\partial}{c\,\partial t} + \frac{\partial x}{c\,\partial t'}\frac{\partial}{\partial x} = \frac{\frac{\partial}{c\,\partial t} + \frac{V}{c}\nabla^x}{\sqrt{1 - V^2/c^2}},$$

$$\nabla' = \left(\frac{\partial x}{\partial x'}\frac{\partial}{\partial x} + \frac{c\,\partial t}{\partial x'}\frac{\partial}{c\,\partial t},\ \frac{\partial}{\partial y'},\ \frac{\partial}{\partial z'}\right) = \left(\frac{\nabla^x + \frac{V}{c}\frac{\partial}{c\,\partial t}}{\sqrt{1 - V^2/c^2}},\ \nabla^y,\ \nabla^z\right).$$

$$(10.18)$$

Notice that these operators have the Lorentz transformation properties of the following four-vector operator:

$$\nabla^\mu \equiv \left(\frac{\partial}{c\,\partial t},\ -\nabla\right). \qquad (10.19)$$

Indeed, ∇^μ is a four-vector known as the *four-gradient* ∇^μ.

Let us return to Maxwell's equations (10.1). By taking the time derivative of the first one and substituting into the fourth one with the divergence applied to both sides we obtain the continuity equation for the electric charge density:

$$\frac{\partial \varrho}{\partial t} + \nabla \cdot j = 0, \qquad (10.20)$$

which expresses the local conservation law of the electric charge. Imposing the principle of relativity means that this equation should be satisfied for all inertial observers. Therefore, the form of equation (10.20) should remain unaffected in a moving, primed frame of reference. This will be the case provided that $(c\varrho, j)$ forms a four-vector j^μ, because then the left-hand side of (10.20) is a product of two four-vectors, which is a relativistic invariant:

$$\eta_{\mu\nu}\nabla^\mu j^\nu = 0. \qquad (10.21)$$

This newly defined four-vector $j^\mu \equiv (c\varrho, j)$ is called the *four-current*. The fact that j^μ is a four-vector imposes definite transformation properties for electric charge density ϱ and electric current density j, but it also guarantees that the charge is locally conserved in all inertial frames.

Let us now rewrite Maxwell's equations (10.1) using the scalar and vector potentials, φ and A. Any vector field with a vanishing divergence can be expressed as a rotation of another vector field. From the second equation (10.1) it follows that we can write:

$$B \equiv \nabla \times A. \tag{10.22}$$

By substituting this into the third equation (10.1) we obtain

$$\nabla \times \left(E + \frac{\partial A}{\partial t} \right) = 0. \tag{10.23}$$

Another mathematical theorem states that any vector field with a vanishing rotation can be expressed as a gradient of a scalar field. In electrostatics, when the rotation of the electric field vanishes, that field can be expressed as a gradient of the scalar potential, $E \equiv -\nabla\varphi$. In a general, dynamical scenario given by (10.23) we have:

$$E \equiv -\nabla\varphi - \frac{\partial A}{\partial t}. \tag{10.24}$$

By substituting (10.24) into the remaining pair of Maxwell's equations (the first and the fourth) (10.1) we obtain

$$-\Delta\varphi - \frac{\partial}{\partial t}\nabla \cdot A = \frac{\varrho}{\varepsilon_0}, \tag{10.25}$$

$$\nabla \times (\nabla \times A) = \nabla(\nabla \cdot A) - \Delta A = \mu_0 j - \frac{1}{c^2}\left(\frac{\partial^2 A}{\partial t^2} + \frac{\partial}{\partial t}\nabla\varphi \right).$$

Both equations look terrible. Instead of nice simplicity we have too many nasty terms. However, the simplicity can be easily restored. Both scalar and vector potentials are not uniquely defined; we have a certain freedom of their choice, known as the *gauge freedom*. The vector potential A is defined up to an arbitrary divergence, because a rotation of divergence vanishes identically. Therefore, we will choose the divergence of the vector potential in such a way that some of the terms in (10.25) cancel out. Our particular choice of that divergence, called the *Lorenz gauge*,[a] has the form:

$$\frac{1}{c^2}\frac{\partial \varphi}{\partial t} + \mathbf{\nabla} \cdot \mathbf{A} = 0. \tag{10.26}$$

With such a choice both the Maxwell's equations (10.25) simplify beautifully:

$$\left(\frac{1}{c^2}\frac{\partial^2}{\partial t^2} - \Delta\right)\varphi = \frac{\varrho}{\varepsilon_0},$$
$$\left(\frac{1}{c^2}\frac{\partial^2}{\partial t^2} - \Delta\right)\mathbf{A} = \frac{\mathbf{j}}{c^2\varepsilon_0}. \tag{10.27}$$

Our efforts start to pay off. But the best part is still to come. Notice that the differential operator appearing in the left-hand side of both of the equations is simply a square of the four-gradient: $\frac{1}{c^2}\frac{\partial^2}{\partial t^2} - \Delta = \eta_{\mu\nu}\nabla^\mu\nabla^\nu \equiv \Box$. As such, it is a relativistic invariant. The right-hand sides of the equations correspond to the components of the four-current j^μ, therefore, for the entirety of the equations to be relativistically invariant, the left-hand sides must also be components of a four-vector. We will denote that four-vector as $A^\mu \equiv (\frac{\varphi}{c}, \mathbf{A})$ and call it *four-potential*. Using the new notation the remaining pair of the Maxwell's equations (10.27) can be written as

$$\Box A^\mu = \mu_0 j^\mu. \tag{10.28}$$

Together with the Lorenz gauge condition (10.26) that can be written as $\eta_{\mu\nu}\nabla^\mu A^\nu = 0$ and the definition of electric and magnetic fields via their potentials, equation (10.28) is fully equivalent to the full set of Maxwell's equations (10.1) with the principle of relativity included. We have finally arrived at a manifestly relativistic formulation of the classical theory of electrodynamics.

10.4   Field Transformations

The four-vector structure of A^μ imposes how the scalar and vector potentials transform between inertial frames of reference.

---

[a]The name Lorenz is often confused with Lorentz, the grandfather of the Lorentz transformation.

For motion with the velocity V along the x axis they undergo a standard Lorentz transformation (1.7):

$$\varphi'(r', t') = \frac{\varphi(r, t) - VA^x(r, t)}{\sqrt{1 - V^2/c^2}},$$

$$A'^x(r', t') = \frac{A^x(r, t) - V\varphi(r, t)/c^2}{\sqrt{1 - V^2/c^2}}.$$

(10.29)

Notice, that in order to complete the transformation we need to apply the formulas both to the components of the four-potential and to their spacetime arguments (ct, r). The newly discovered four-vector also explains the mysterious Lorentz factor present in the denominator of the expression (10.15).

Our next step is to determine the transformation laws for the electric and magnetic fields, E and B. All we need to do is to substitute the equations (10.29) into the definitions (10.22) and (10.24), which yields:

$$E'^x = E^x, \qquad\qquad B'^x = B^x,$$

$$E'^y = \frac{E^y - VB^z}{\sqrt{1 - V^2/c^2}}, \qquad B'^y = \frac{B^y + VE^z/c^2}{\sqrt{1 - V^2/c^2}}, \qquad (10.30)$$

$$E'^z = \frac{E^z + VB^y}{\sqrt{1 - V^2/c^2}}, \qquad B'^z = \frac{B^z - VE^y/c^2}{\sqrt{1 - V^2/c^2}},$$

where for simplicity we have dropped the dependence of the fields on the spacetime coordinates (it is the same as in the equations (10.29)). The above transformation law can be generalised to an arbitrary velocity V:

$$E' = \frac{E + V \times B}{\sqrt{1 - V^2/c^2}} - (E \cdot V)\frac{V}{V^2}\left(\frac{1}{\sqrt{1 - V^2/c^2}} - 1\right),$$

$$B' = \frac{B - V \times E/c^2}{\sqrt{1 - V^2/c^2}} - (B \cdot V)\frac{V}{V^2}\left(\frac{1}{\sqrt{1 - V^2/c^2}} - 1\right).$$

(10.31)

Notice that in the non-relativistic limit of $c \to \infty$ the formulas reduce to the following equations:

$$E' = E + V \times B,$$
$$B' = B. \tag{10.32}$$

So, if the magnetic field vanishes in one reference frame, it also vanishes in all other frames. And the second equation (10.27) shows that when $c \to \infty$, the electric current j does not give rise to the creation of a magnetic field. It looks as though the whole magnetic field was just a relativistic correction to the electric field. So, in principle, the existence of magnets proves that the speed of light, c, is finite, just like we have foreshown in Section 2.5, when we used the magnetic component of the Lorentz force to prove the realness of the Lorentz contraction.

It is clear from the transformation equations (10.30) that only the component of the field that is perpendicular to the direction of motion gets affected by the transformation. However, we also have to take into account that the spacetime coordinates are Lorentz contracted along the direction of motion, and not perpendicular to the velocity.

To get a better picture, let us consider a simple example of a point-like charge q moving with the velocity v along x. The corresponding electric field can be found by transforming the electric Coulomb field of a resting charge to a frame moving with the velocity $-v$, which leads to

$$E'^{x}(r', t') = \frac{q\,x}{4\pi\varepsilon_0(x^2 + y^2 + z^2)^{3/2}},$$

$$E'^{y}(r', t') = \frac{q\,y}{4\pi\varepsilon_0(x^2 + y^2 + z^2)^{3/2}\sqrt{1 - v^2/c^2}}, \tag{10.33}$$

$$E'^{z}(r', t') = \frac{q\,z}{4\pi\varepsilon_0(x^2 + y^2 + z^2)^{3/2}\sqrt{1 - v^2/c^2}},$$

where the unprimed quantities correspond to the rest frame of the charge, and the primed ones are in a frame in which the charge is moving with the velocity v. Notice the following relation:

$$\frac{E'^x}{E'^y} = \frac{x}{y}\sqrt{1 - v^2/c^2} = \frac{x' - vt'}{y'}, \tag{10.34}$$

showing that the field is still radially orientated with respect to the moving charge. Moreover, the field ahead and behind the moving charge is weaker than the field on the sides:

$$E'^x(x', 0, 0, 0) = \frac{q}{4\pi\varepsilon_0 x'^2}\left(1 - \frac{v^2}{c^2}\right),$$

$$E'^y(0, y', 0, 0) = \frac{q}{4\pi\varepsilon_0 y'^2}\frac{1}{\sqrt{1 - v^2/c^2}}. \tag{10.35}$$

It appears as if the space containing the lines of the electric field was Lorentz contracted — see Fig. 10.2. We have to remember of course, that in the moving frame the additional magnetic component of the field emerges. That new field circulates "around" the trajectory of the moving charge, similar to the magnetic field circulating around the electric current.

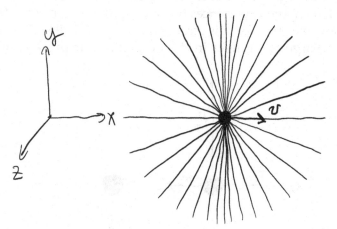

Figure 10.2: Lines of the electric field of a charge moving at a constant speed along x.

10.5   Lorentz Force and Newton's Third Law

We gained some basic knowledge about electromagnetic fields. To make things complete we should also discuss what these fields actually do and how they affect electric charges. It should not come as a surprise that the Newton's second law for the fields E and B takes the form:

$$\frac{\mathrm{d}p}{\mathrm{d}t} = q(E + v \times B), \qquad (10.36)$$

where v is the velocity of the charge in the considered reference frame. This is the so-called *Lorentz force* and it already has the correct relativistic form.

But here is an intriguing example. Consider a pair of identical point-like charges q moving with equal, constant speeds along two straight lines as shown in Fig. 10.3 [22]. At this instant, the charge moving vertically is crossing the trajectory of the charge moving horizontally. What are the mutual forces acting on both charges? The answer is rather straightforward, because we know the Lorentz force formula (10.36) and the electromagnetic fields surrounding the moving charges, but it will surprise us nonetheless.

The force acting on the right charge is purely electrical, because the magnetic field created by the left charge at its own trajectory vanishes. In Fig. 10.3, that repelling electrical force is denoted as F_1. The left charge experiences the analogous electric force denoted as F_2, however this force is stronger, because the electric field of a moving charge is stronger in the perpendicular plane than along

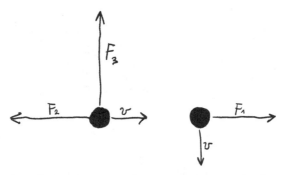

Figure 10.3:   Mutual electromagnetic forces acting on a pair of moving charges. Will this scenario violate the third Newton's law?

the trajectory — see the formulas (10.35). Moreover, there appears to be an additional magnetic force F_3 acting on the left charge. It is directed perpendicular to both F_1 and F_2. Does it all mean that Newton's third law fails in relativistic settings and the mutual forces are not equal? If that was the case, it would mean that the total momentum of the system containing interacting relativistic particles was not conserved. And that is something that we definitely would like to avoid. Unfortunately, we will need to wait until Section 10.6 for the solution to this problem, when we learn a bit more about the properties of electromagnetic fields.

10.6 The Energy and Momentum of Fields

It should not come as a surprise that electromagnetic fields carry energy. As a consequence, we should also expect that they carry momentum as well. What is the electrodynamical energy and momentum of the fields? To figure it out, let us manipulate Maxwell's equations (10.1) a little. By scalar multiplying the third equation by B, and the fourth one by E, and then subtracting the resulting equations, we obtain

$$E \cdot (\nabla \times B) - B \cdot (\nabla \times E) = \mu_0 E \cdot j + \mu_0 \varepsilon_0 E \cdot \frac{\partial E}{\partial t} + B \cdot \frac{\partial B}{\partial t}. \quad (10.37)$$

The left-hand side is simply equal to $-\nabla \cdot (E \times B)$, and the last two terms on the right-hand side can be cast in a different form, namely: $E \cdot \frac{\partial E}{\partial t} = \frac{\partial}{\partial t} \frac{E^2}{2}$ and $B \cdot \frac{\partial B}{\partial t} = \frac{\partial}{\partial t} \frac{B^2}{2}$, therefore, after dividing both sides by μ_0 and doing some minor reshuffling, we can rewrite (10.37) in the following form:

$$\frac{\partial}{\partial t}\left(\frac{\varepsilon_0 E^2}{2} + \frac{B^2}{2\mu_0}\right) + \nabla \cdot \left(\frac{1}{\mu_0} E \times B\right) = -E \cdot j. \quad (10.38)$$

The expression on the right-hand side, $-E \cdot j$, is the work of the field upon the material conducting the electric current density j, taken with a negative sign. In other words, it is the electromagnetic energy transferred into matter through heat. In the case of a vacuum, when no conductors or currents are present, equation (10.38) becomes a continuity equation of the form (10.20) expressing the law of local conservation of energy of the electromagnetic field:

$$\frac{\partial \varrho_E}{\partial t} + \nabla \cdot S = 0, \tag{10.39}$$

where the quantities:

$$\varrho_E = \frac{\varepsilon_0 E^2}{2} + \frac{B^2}{2\mu_0},$$

$$S = \frac{1}{\mu_0} E \times B \tag{10.40}$$

correspond to the local energy density of the electromagnetic field and the density of the energy current (or the density of the momentum of the field), known as the *Poynting vector*. Notice that ϱ_E and S do not form a four-vector unlike the charge density and the current density $(\varrho c, j)$, because (10.39) is only satisfied in the special case of the vacuum; in general we have (10.38).

We can also address the problem from Section 10.5, depicted in Fig. 10.3, in which the total momentum did not seem to be conserved. Notice that the total momentum of the system should also involve the momentum of the electromagnetic field. That field momentum is given by the Poynting vector defined by the second equation (10.40), which is a nonlinear function of fields E and B. Because of this nonlinearity, the field momentum from the pair of charges in the example shown in Fig. 10.3 is not the sum of the individual momenta from each charge. Moreover, when the charges move, the total momentum of the field changes. This is where we should seek the missing momentum of the system. The mechanical momentum of the particles may not be conserved, but when we take into account the changing momentum of the field involving the Poynting vector, the total momentum should remain constant.

The Poynting theory of field momentum solves some questions, but brings disturbing answers. Consider a pair of opposite resting charges creating some static electric field around them — see Fig. 10.4(a). That field carries a certain energy that must move around when we move the charges. Imagine that we start to move

(a) (b)

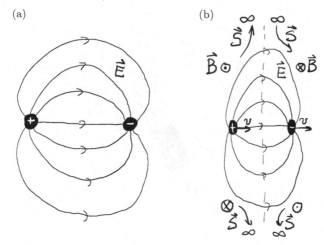

Figure 10.4: Two opposite electric charges (a) at rest; (b) moving horizontally with a constant speed.

the system of charges horizontally, as shown in Fig. 10.4(b). Due to Lorentz contraction the horizontal distances will shrink affecting the electric field. Also, a new magnetic field circling around the moving charges will appear as shown in the figure. Let us ask a simple question: which way does the field energy travel from left to right? The answer can be given by calculating the Poynting vector field S defined in the second formula (10.40). It turns out that the energy does not simply flow from the left to the right. At the dashed plane of symmetry shown in Fig. 10.4(b), the electric field E is horizontal and the magnetic field vanishes, therefore there can be no Poynting vector going through that surface! So, how does the energy travel from left to right? It flows through infinity. On the left-hand side of the dashed plane of symmetry it flows away to infinity and on the other side it returns from infinity. The local energy of the field is obviously conserved due to (10.39), but the energy flow is quite peculiar.

Another strange example involves a static magnet with a static charge placed inside, as shown in Fig. 10.5. Both the electric and magnetic fields are static, but there exists a non-trivial Poynting vector field constantly circling around the system, suggesting a stationary energy flow within the static electromagnetic field.

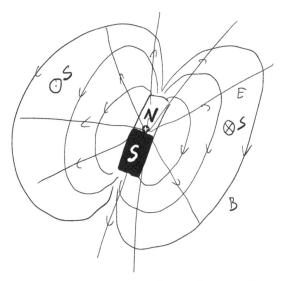

Figure 10.5: A static magnet and a static electric charge produce a static electro-magnetic field with the non-vanishing Poynting vector field.

10.7 Electrodynamics vs the Principle of Equivalence

We have saved the most challenging, but also the most intriguing, problem for last. Suppose that someone hung an electric charge on a rope in a gravitational field, as shown in Fig. 10.6. According to the principle of equivalence, the effect of gravity on the charge should be indistinguishable from the effect of uniform acceleration. But accelerated charges radiate energy away. So, is the hanging charge going to radiate the energy away as well? If so, what is the source of that energy? If not, what is going wrong with the principle of equivalence?

Bondi, Born, Feynman, Pauli, Peierls, Sommerfeld — these are only a few of the famous physicists who have attempted to answer this question. Each of them provided a yes-or-no answer and, interestingly, all these answers were mutually exclusive. This means that this puzzle is rather difficult, but we will try to address it anyway.

We will begin by investigating the scenario of a uniformly accelerated, charged particle, and then we will try to argue what happens in the case of the "real" gravitational field. Let us first

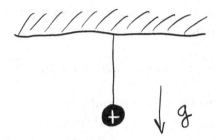

Figure 10.6: An electric charge hanging on a rope in a gravitational field. Is it going to radiate the energy away?

derive the Liénard–Wiechert potentials (10.7) and (10.8) for a particle of a charge q, moving along a uniformly accelerated, hyperbolic trajectory with the proper acceleration a. It is convenient to parametrise the trajectory with its proper time τ by substituting $\chi = \frac{c^2}{a}$ into equations (8.19):

$$ct(\tau) = \frac{c^2}{a} \sinh \frac{a\tau}{c},$$
$$x(\tau) = \frac{c^2}{a} \cosh \frac{a\tau}{c}. \tag{10.41}$$

We will be looking for the Liénard–Wiechert potentials φ and A at an arbitrary spacetime point (ct, r). The equation for the retarded time is given by

$$c\,(t - t_{\text{ret}})^2 = (x - x_{\text{ret}})^2 + y^2 + z^2. \tag{10.42}$$

Let us substitute (10.41) for the retarded time and position with $\psi \equiv \frac{a\tau_{\text{ret}}}{c}$ into (10.42), and parametrise the observation point by $ct \equiv \rho \sinh \phi$ and $x \equiv \rho \cosh \phi$, yielding:

$$\left(\rho \sinh \phi - \frac{c^2}{a} \sinh \psi \right)^2 = \left(\rho \cosh \phi - \frac{c^2}{a} \cosh \psi \right)^2 + y^2 + z^2. \tag{10.43}$$

By squaring the brackets, reshuffling some terms and using the hyperbolic identity $\cosh(x - y) = \cosh x \cosh y - \sinh x \sinh y$, we obtain the expression:

$$\cosh(\psi - \phi) = \frac{\rho^2 + y^2 + z^2 + (c^2/a)^2}{2\rho c^2/a}. \tag{10.44}$$

In principle, the retarded proper time $\tau_{\text{ret}} = \frac{c}{a}\psi$ can be obtained by inverting the cosh function, but for now we will keep our result as-is. In order to derive the Liénard–Wiechert potentials, we need to calculate the following retarded expression:

$$\left(R - \frac{\mathbf{R} \cdot \mathbf{v}}{c}\right)_{\text{ret}} = c(t - t_{\text{ret}}) - (x - x_{\text{ret}})\frac{v_{\text{ret}}}{c}. \tag{10.45}$$

Let us substitute to the above equation $ct = \rho \sinh \phi$, $x = \rho \cosh \phi$, $ct_{\text{ret}} = \frac{c^2}{a} \sinh \psi$, $x_{\text{ret}} = \frac{c^2}{a} \cosh \psi$ and compute the retarded velocity of the particle: $v_{\text{ret}} = \frac{dx}{dt} = c \operatorname{tgh} \psi$. Finally, let us apply the identity $\sinh(x - y) = \sinh x \cosh y - \cosh x \sinh y$. This procedure yields:

$$\left(R - \frac{\mathbf{R} \cdot \mathbf{v}}{c}\right)_{\text{ret}} = \rho \frac{\sinh(\phi - \psi)}{\cosh \psi}. \tag{10.46}$$

Our newly derived formula gives us the final form of the Liénard–Wiechert potentials for the uniformly accelerated charge q:

$$\begin{aligned} \varphi(r, t) &= -\frac{q \cosh \psi}{4\pi\varepsilon_0 \rho \sinh(\psi - \phi)}, \\ A(r, t) &= \frac{qc \sinh \psi}{4\pi\varepsilon_0 \rho \sinh(\psi - \phi)}\hat{x}, \end{aligned} \tag{10.47}$$

where $\hat{x}$ is a unit vector along the direction of motion and ψ is given by (10.44). We are now ready to address the main problem. The obtained results are rather complicated, but notice how the vector potential A simplifies for $t = 0$, which corresponds to $\phi = 0$. In this special case the dependence on ψ cancels out and the vector potential (10.47) reduces to:

$$A(r, 0) = \frac{qc}{4\pi\varepsilon_0 x}\hat{x}. \tag{10.48}$$

The corresponding magnetic field $B = \nabla \times A = 0$ vanishes, as does the Poynting vector $S = \frac{1}{\mu_0}E \times B = 0$ at $t = 0$. There is therefore, no radiation at all in the inertial frame instantaneously co-moving with the accelerated charge.

We have previously constructed a non-inertial frame based on the principle that observations carried out in this frame coincide with the observations of the instantaneously co-moving inertial observer. Therefore, we conclude that there is no magnetic field and there is no Poynting vector in the uniformly accelerated reference frame co-moving with the accelerated charge at $t = 0$. But since no instant of time is preferred in that frame, whatever happens at $t = 0$ must also happen at any other time. This perspective should also be equivalent to the perspective of a guy watching the electric charge hanging on a rope in a gravitational field, therefore there can be no radiation in this case either. The principle of equivalence is saved, brilliant!

But we are not done yet; the most interesting bit is still waiting to be discovered. It turns out that, although radiation is absent in the uniformly accelerated frame co-moving with the charge, it is present for the inertial observer. Here is why. Let us first derive the expressions for the electromagnetic fields resulting from the potentials (10.47). The electric field is given by $E = -\nabla\varphi - \frac{\partial A}{\partial t}$, and after some nasty algebra it turns out to be [23]:

$$E^x = \frac{q}{4\pi\varepsilon_0 \varrho \sinh^2(\psi - \phi)}\frac{\partial\psi}{\partial\varrho} = \frac{qa}{4\pi\varepsilon_0 c^2 \varrho^2}\frac{\varrho - \frac{c^2}{a}\cosh(\psi - \phi)}{\sinh^3(\psi - \phi)},$$

$$E^y = \frac{q\cosh\phi}{4\pi\varepsilon_0 \varrho \sinh^2(\psi - \phi)}\frac{\partial\psi}{\partial y} = \frac{qa}{4\pi\varepsilon_0 c^2 \varrho^2}\frac{y\cosh\phi}{\sinh^3(\psi - \phi)}, \quad (10.49)$$

$$E^z = \frac{q\cosh\phi}{4\pi\varepsilon_0 \varrho \sinh^2(\psi - \phi)}\frac{\partial\psi}{\partial z} = \frac{qa}{4\pi\varepsilon_0 c^2 \varrho^2}\frac{z\cosh\phi}{\sinh^3(\psi - \phi)}.$$

Similarly, all the components of the magnetic field $B = \nabla \times A$ are equal to

$$B^x = 0,$$

$$B^y = -\frac{q\sinh\phi}{4\pi\varepsilon_0 c \varrho \sinh^2(\psi - \phi)}\frac{\partial\psi}{\partial z} = \frac{qa}{4\pi\varepsilon_0 c^3 \varrho^2}\frac{z\sinh\phi}{\sinh^3(\psi - \phi)}, \quad (10.50)$$

$$B^z = \frac{q\sinh\phi}{4\pi\varepsilon_0 c \varrho \sinh^2(\psi - \phi)}\frac{\partial\psi}{\partial y} = -\frac{qa}{4\pi\varepsilon_0 c^3 \varrho^2}\frac{y\sinh\phi}{\sinh^3(\psi - \phi)}.$$

From these results we can calculate the energy density of the electromagnetic field (10.40). Let us determine what energy density ϱ_E is emitted by the accelerated charge at the retarded moment $\psi = 0$, when the charge was instantaneously at rest:

$$\varrho_E = \frac{q^2 a^2}{32\pi^2 \varepsilon_0 c^4 \varrho^4 \sinh^6 \phi}$$

$$\times \left[\left(\varrho - \frac{c^2}{a}\cosh\phi\right)^2 + \left(y^2 + z^2\right) \times (1 + 2\sinh^2\phi)\right]. \quad (10.51)$$

At some later observation time t that emitted radiation will reach a sphere of radius $R = ct$, centred around the initial position of the charge $\left(\frac{c^2}{a}, 0, 0\right)$:

$$c^2 t^2 = \left(x - \frac{c^2}{a}\right)^2 + y^2 + z^2. \quad (10.52)$$

After plugging the definitions $ct = \varrho\sinh\phi$ and $x = \varrho\cosh\phi$ into (10.52) and after some rearrangements we obtain:

$$\frac{c^4}{a^2}\sinh^2\phi = \left(\varrho - \frac{c^2}{a}\cosh\phi\right)^2 + y^2 + z^2 \quad (10.53)$$

and by substituting this result into (10.51) we arrive at

$$\varrho_E = \frac{q^2}{32\pi^2 \varepsilon_0 R^4} + \frac{q^2 a^2}{16\pi^2 \varepsilon_0 c^4 R^4}\left(y^2 + z^2\right). \quad (10.54)$$

The first term does not depend on acceleration and corresponds to the energy density of the Coulomb field of a charge at rest at the distance R. The second term is more interesting, as it depends on the proper acceleration a, and falls off with a distance like $\sim \frac{1}{R^2}$, therefore, it clearly corresponds to some sort of electromagnetic radiation. The density of that radiation is at its maximum in the sideways direction, and vanishes along the trajectory of that source. The amount of energy emitted by the accelerating charge within an infinitesimal time dt is equal to

$$dE_{rad} = \int_{sphere} \varrho_E \, c \, dt \, R^2 d\Omega, \quad (10.55)$$

where cdt is the infinitesimal thickness of the shell containing the radiation spreading with the speed of light, and $d\Omega$ is the solid angle. In the limit of $R \to \infty$, i.e. far away from the accelerating charge, the Coulomb term in (10.54) becomes irrelevant and vanishes, while the radiation term stays constant. The integration in (10.55) involving the radiation term in (10.54) is straightforward if we use the symmetry trick:

$$\int_{\text{sphere}} \left(y^2 + z^2\right) d\Omega = \frac{2}{3} \int_{\text{sphere}} \left(x^2 + y^2 + z^2\right) d\Omega = \frac{2}{3} 4\pi R^2.$$
(10.56)

Finally, the power of radiation from the accelerating charge at instantaneous rest (as well as for small, non-relativistic velocities) takes the form of the *Larmor formula*:

$$\frac{dE_{\text{rad}}}{dt} = \frac{q^2 a^2}{16\pi^2 \varepsilon_0 c^4 R^4} \frac{2}{3} c R^4 4\pi = \frac{q^2 a^2}{6\pi \varepsilon_0 c^3}.$$
(10.57)

Let us summarise. We have just shown that a uniformly accelerated charge radiates energy proportional to the square of the proper acceleration in the inertial frame of reference, but does not radiate energy in the co-moving, uniformly accelerating frame. This bizarre conclusion was first reached by Boulware in [24, 25]. Therefore, his (and our) answer to the long-standing question of whether an electric charge hanging in a gravitational field (or uniformly accelerating) radiates energy is the truly relativistic: "it depends".

10.8 Epilogue

There is an interesting epilogue to this story. In quantum theory, classical electrodynamical radiation is replaced by a stream of quantum particles of light — photons. We have briefly discussed them in Chapter 4 and we have learned that they have a very peculiar nature. They behave in a non-deterministic, unpredictable way, they have the ability of being present at several places at the same time, and due to the Doppler effect their colour depends on the observer. But if Boulware's conclusion about the relativity of radiation is correct, then we should expect that the presence or absence of photons is also relative! And guess what, this is exactly what we learn from quantum field theory in non-inertial frames of reference! The relativity of the existence of quantum particles is known as the *Unruh effect*. Indeed, if a resting observer is placed in a quantum vacuum, then a uniformly accelerated observer in the same vacuum state of the quantum field will witness particles surrounding himself. This effect is also deeply connected with the so-called *Hawking radiation* of black holes. Sounds interesting, but that is a topic for a whole different story.

10.9 Questions

- Why must a four-current be a four-vector?
- Why must a four-potential be a four-vector?
- Is the Lorenz gauge condition the same in all inertial reference frames?
- Is the Poynting vector a component of some four-vector?
- Which way do electric wires heat up when conducting electric current? Use the Poynting theory to determine the stream of energy of the field that converts into heat.
- Can you propose an experiment testing the idea that a magnet combined with the static electric charge shown in Fig. 10.5 generates some circulating energy flow?
- Does an electric charge hanging on a rope in a gravitational field radiate the electromagnetic energy away?

10.10 Exercises

- An infinitely long straight electrical wire of a cross-section S is electrically uncharged and conducts the current I in its rest frame. Calculate the electromagnetic field created by this conductor, and Lorentz transform it to the frame that moves with velocity V along the wire. Show that the same result can be obtained by transforming the field sources (currents and charges) first and computing from them the electromagnetic field in the new frame.
- An infinitely long, straight, electrical wire of cross-section S is electrically uncharged and conducts the current I in its rest frame. At distance d from the wire, there is a charge q moving away with

velocity V, perpendicular to the wire. Calculate the Lorentz force acting on the charge and transform it into the rest frame of that charge. Can you justify the presence of the force in the second frame without doing any calculations?

- In some inertial frame of reference the constant fields E and B are not perpendicular. Is there another inertial frame in which these fields become perpendicular? If the answer is affirmative, find that frame.

- In some inertial frame of reference the constant fields E and B are not perpendicular. Is there another inertial frame, in which these fields become parallel? If the answer is affirmative, find that frame.

- Determine the transformation properties of the scalar product: $E \cdot B$.

- Determine the transformation properties of the following expression: $E \cdot E - c^2 B \cdot B$.

- Derive the transformation laws (10.31) from the special case (10.29).

- Consider the example of two repelling charges depicted in Fig. 10.3. Verify if at the instant presented in the figure, the total momentum of the system involving both the charges and the fields is conserved.

- A field line of the electric field shown in Fig. 10.7 begins at some positive charge and enters another, negative charge. Assuming that there are no other charges around, find the relation between an angle α at which the line leaves the first charge and the angle β at which it enters the other, relative to the straight line connecting the charges.

- Suppose that the density of lines of the electric field represents the local intensity of the field. Use the formulas (10.35)

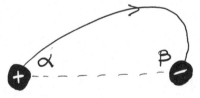

Figure 10.7:   A line of an electric field leaving a positive charge at an angle α and entering a negative charge at an angle β.

to verify the idea that the electric lines of a moving charge can be obtained by Lorentz contracting the electric lines of a resting charge.

- Do the Liénard–Wiechert potentials (10.47) satisfy Maxwell's equations at the white hole event horizon $x = -ct$? How can you interpret the result and, if something seems "off" about it, how can it be fixed?

Bibliography

[1] L. D. Landau and E. M. Litshitz, *Classical Theory of Fields* (Oxford: Elsevier, 1987).

[2] J. Hafele and R. Keating, Around-the-world atomic clocks: Observed relativistic time gains, *Science* **177**, 4044 (1972).

[3] J. Bailey *et al.*, Measurements of relativistic time dilatation for positive and negative muons in a circular orbit, *Nature* **268**, 301 (1977).

[4] J. Bailey *et al.*, Direct test of relativistic time dilation, *Nucl. Phys.* B **150**, 1 (1979).

[5] K. Lorek, J. Louko, and A. Dragan, Ideal clocks — A convenient fiction, *Class. Quantum Grav.* **32**, 175003 (2015).

[6] R. Pierini, K. Turzyński, and A. Dragan, Can a charged decaying particle serve as an ideal clock in the presence of the magnetic field?, *Phys. Rev. D* **97**, 045006 (2018).

[7] L. H. Thomas, The motion of the spinning electron, *Nature* **117**, 514 (1926).

[8] A. Dragan and T. Odrzygóźdź, Half-page derivation of the Thomas precession, *Am. J. Phys.* **81**, 631 (2013).

[9] P. Lewulis and A. Dragan, Three-line derivation of the Thomas precession, *Am. J. Phys.* **87**, 674 (2019).

[10] W. Ignatowsky, Das Relativitätsprinzip (Principle of Relativity), *Arch. Math. Phys.* **17**, (1910); P. Frank and H. Rothe, Über die Transformation der Raumzeitkoordinaten von ruhenden auf bewegte Systeme, *Ann. der Phys.* **34**, 825 (1911).

[11] A. Szymacha, *Przestrzeń i ruch* (Space and Motion), (Wydawnictwo Uniwersytetu Warszawskiego, 1997).

[12] A. Dragan and A. Ekert, Quantum principle of relativity, *New J. Phys.* **22**, 033038 (2020).

[13] L. Machildon, A. Antippa, and A. Everett, Superluminal coordinate transformations: The two-dimensional case, *Can. J. Phys.* **61**, 256 (1983).

[14] L. Machildon, A. Antippa, and A. Everett, Superluminal coordinate transformations: Four-dimensional case, *Phys. Rev. D* **27**, 1740 (1983).

[15] R. Sutherland and J. Shepanski, Superluminal reference frames and generalized Lorentz transformations, *Phys. Rev. D* **33**, 2896 (1986).

[16] W. Rindler, Length contraction paradox, *Am. J. Phys.* **29**, 365 (1961).

[17] G. Gamov, *Mr Tompkins in Wonderland* (New York, NY: Macmillan Company, 1939).

[18] A. Nowojewski, J. Kallas, and A. Dragan, On the appearance of moving bodies, *Amer. Math. Month.* **111**, 817 (2004).

[19] M. L. Boas, Apparent shape of large objects at relativistic speeds, *Am. J. Phys.* **5**, 283 (1961).

[20] G. D. Scott and H. J. Van Driel, Geometrical appearances at relativistic speeds, *Am. J. Phys.* **38**, 971 (1970).

[21] A. Szymacha, A public talk delivered at the University of Warsaw (2007).

[22] R. P. Feynman, R. B. Leighton, and M. Sands, *Feynman's Lectures in Physics*, vol. II (Redwood City, CA: Addison–Wesley, 1964).

[23] W. Pauli, *Theory of Relativity* (Dover Publications, New York, 1958).

[24] D. Boulware, Radiation from a uniformly accelerated charge, *Ann. Phys.* **124**, 169 (1980).

[25] R. Peierls, *Surprises in Theoretical Physics* (Princeton, NJ: Princeton University Press, 1979).

[26] S. Fulling, Nonuniqueness of canonical field quantization in Riemannian spacetime, *Phys. Rev. D* **7**, 2850 (1973); P. Davies, Scalar production in Schwarzschild and Rindler metrics, *J. Phys. A* **8**, 609 (1975); W. Unruh, Notes on black-hole evaporation, *Phys. Rev. D.* **14**, 870 (1976).

Index

Printed in the United States
by Baker & Taylor Publisher Services